Methodologies for the Study of Low-Level Radiation in the Midwest

by

Charles W. Huver, Gertrude A. Dixon, Naomi Jacobson, George I. J. Dixon.

ANVIL PRESS

ACKNOWLEDGMENTS

We are indebted to the many scientists who have pioneered investigations of radioactive fallout, its pathways and its biological effects on humankind. Without their factual groundwork our methodological studies would not have been possible. Their many contributions are suggested by our bibliographies.

These researchers and the factual background they contribute share acknowledgment for our studies with the many humanely concerned citizens who asked the many unanswered questions about radiation, some of which we have herein attempted to answer.

The Wisconsin researchers owe a special debt of gratitude to the late Dorothy Jones, without whose inspiration and persistant encouragement the study would never have been undertaken, and to Larry Bogart, director of the Citizens Energy Council, who has consistently nourished the efforts of citizen researchers. For helpful critical commentary we are indebted to Dr. Ernest J. Sternglass, Dr. Thomas F. Mancuso, Dr. Rosalie Bertell, Dr. L. Douglas DeNike, Dr. Gerald Drake and Martha Drake. To Dr. John W. Gofman – tireless researcher, generous teacher and thorough critic – we are immeasurably grateful.

Many midwestern citizens and organizations gave generously to the financial support of these studies, including the Environmental Council, University of Wisconsin-Stevens Point and the Midwest Environmental Education and Research Association.

The inhabitants of our Earth entered the nuclear age blindfolded. We hope our efforts to draw together existing information and suggest further radiation research directions will help to remove some of those blindfolds, so that we might all with clearer vision foresee our nuclear future.

The Authors

DEDICATED TO THE MEMORY
OF CITIZEN RESEARCHERS:

Elise Jerard
Karen Silkwood
Dorothy Jones
Paul Jacobs

Manufactured in U. S. A.
First printing November 1979
Library of Congress No. 79-56790

ISBN: 0-918552-07-9 (paper)
ISBN: 0-918552-08-7 (cloth)

Printed at the Anvil Press shop in Millville, Minn.

A Plan for the Study of Low-Level Radiation Health Effects in Minnesota

by

Charles W. Huver, Ph.D.

A PLAN FOR THE STUDY OF LOW-LEVEL RADIATION HEALTH EFFECTS IN MINNESOTA

The pathologic symptoms of radiation sickness and the carcinogenic action of ionizing radiation have been recognized for more than half a century. Radiation in large doses is widely acknowledged to produce serious somatic, genetic, and carcinogenic effects, but the biological significance of low doses remains clouded in controversy. The United Nations Scientific Committee on the Effects of Atomic Radiation (1964) has expressed concern that even a small increase in environmental radiation may magnify the risk of cancer. The question of the biological and public health consequences of low dose exposure becomes increasingly important as a result of expanding uses of radiation-emitting equipment and isotopes in modern medicine, a rapidly expanding nuclear power technology, and nuclear weapons testing which has resulted in worldwide radioactive contamination by fallout.

Minnesota is a particularly favorable site for a study of the biological consequences of increases in environmental radioactivity because portions of the state have received unusually heavy levels of fallout primarily from nuclear weapons testing in the atmosphere at the Nevada test site. Other contributions to above background levels have come from ventings and leakages from underground nuclear explosions (Martell, 1964), shallow plowshare explosions such as the Sedan event (Krey and Fried, 1965). and from the operations of the now defunct Elk River reactor and deliberate and accidental discharges from the Monticello and Prairie Island reactors.

A preliminary survey of the data on environmental levels of radioactivity in Minnesota reveals that the northwestern corner of the state has had remarkably high levels of fallout. The Federal Radiation Council (1964) has mapped the distribution of strontium-90 levels in milk (see Figure 1) to show an extensive area in northwestern Minnesota which during the year of highest fallout, 1963, had average levels of above 40 pCi/l. This area included Crookston, Thief River Falls, and Red Lake. The U.S. Geological Survey (1969) measured the gross beta activity of thirty-six (36) North American Rivers in the year 1960-61 and found the highest value of 53 pCi/l was recorded for the Red River of the North at Grand Forks, North Dakota. The sampling program of radioactivity in Minnesota municipal water supplies (see Table 1) by the Minnesota Department of Health (1968, 1969) has confirmed the relatively high gross beta levels in the raw water supplies of Crookston and Hallock, the Red Lake River and the South Fork of Two Rivers, respectively.

Barber (1970) has summarized data collected by the Minnesota Department of Health between 1962 and 1967 which show consistently higher levels of strontium-90 in milk of northern Minnesota as compared to southern Minnesota. However, Barber failed to find a satisfactory explanation for this remarkable difference in his review of the literature on strontium-90 content of Minnesota milk.

The testing of nuclear weapons in the atmosphere by the United States, United Kingdom, and the USSR through 1963 produced a TNT equivalent of 250 megatons (Gofman, 1972). When there is a nuclear test in the megaton range, the gases are so slow to cool that a major portion of the radionuclides enter the stratosphere where they are distributed widely. Fission products with long half-lives drift back into the troposhpere before losing their radioactivity and are deposited in patterns which are strongly influenced by meteorological conditions. This type of fallout requires a long time to drift back to earth so that there are marked changes in activity ratios between certain radioisotopes such as ^{144}Ce and ^{90}Sr with such variables as time after a test and altitude (Drevinsky and Pecci, 1965). For nuclear explosions in the kiloton range, the heat of the fireball is much less so that the fission products do not reach

Table 1. Total beta radioactivity in Minnesota raw and treated water supplies, January-June 1968

Town and water source	Type of water	Average concentration (pCi/liter) Jan	Feb	Mar	Apr	May	Jun
Crookston, Red Lake River	Raw	22	20	14	16	15	17
	Treated	21	18	18	13	16	17
East Grand Forks, Red Lake River	Raw	15 (5)*	17 (4)	18 (4)	13 (5)	13 (5)	15 (4)
	Treated	8 (5)	8 (4)	10 (4)	9 (5)	8 (4)	19 (4)
Eveleth, St. Mary's Lake	Raw	15 (5)	13 (3)	12 (3)	12 (4)	13 (5)	17 (4)
	Treated	12 (5)	11 (3)	12 (5)	14 (4)	16 (5)	13 (4)
Fairmont, Budd Lake	Raw	12 (6)	14 (5)	13 (3)	23 (5)	14 (5)	14 (3)
	Treated	7 (6)	5 (5)	7 (3)	7 (5)	10 (5)	7 (3)
Hallock, Two Rivers South Fork ..	Raw	20 (5)	21 (4)	30 (4)	15 (4)	15 (5)	19 (4)
	Treated	10 (5)	9 (4)	12 (4)	7 (4)	8 (5)	10 (4)
International Falls, Rainy River ..	Raw	9	6	6	9	12	5
	Treated	7	8	7	7	8	10
Minneapolis tap water	Raw						
	Treated	5 (5)	6 (4)	7 (4)	7 (5)	8 (5)	7 (4)
St. Cloud, Mississippi River	Raw	9	8	8	10	10	15
	Treated	5	6	5	7	8	7
St. Paul, Vadnais Chain of Lakes .	Raw	9 (5)	9 (4)	10 (4)	9 (5)	11 (4)	12 (4)
	Treated	7 (5)	5 (4)	7 (4)	7 (5)	6 (4)	8 (4)

* Number in parentheses gives number of samples averaged.

FIGURE 1

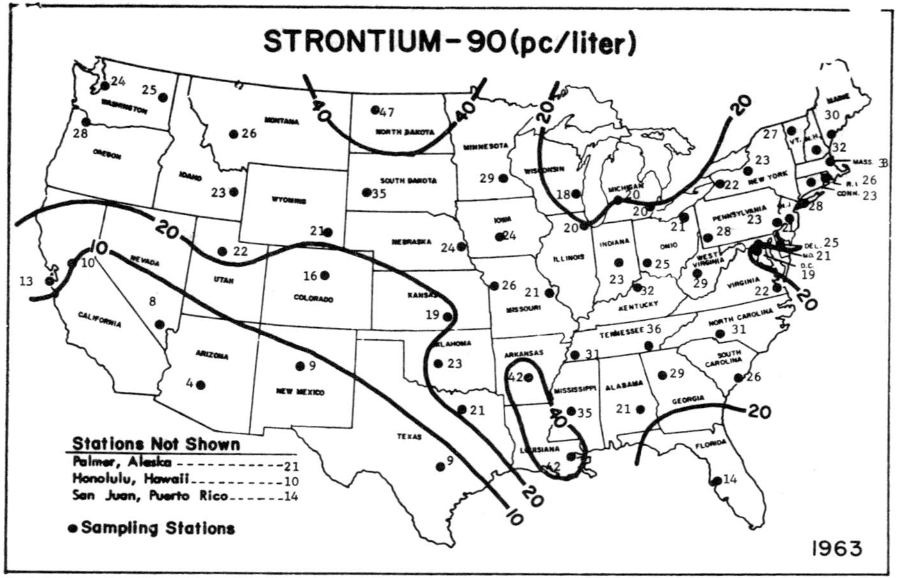

STRONTIUM-90 CONCENTRATIONS IN PASTEURIZED MILK
(Data from Federal Radiation Council Report No. 6, October, 1964)

the stratosphere but remain in the troposphere (FRC, p. 20, 1960).

In the troposphere the cloud is carried by wind action and by turbulent mixing. Changes in temperature and pressure induce both horizontal and vertical movements of the cloud. The cloud becomes progressively depleted of its radioisotope load due to washout by rain, gravitational settling and direct impaction on surfaces. In some twenty to forty days, about one-half of the radioactive debris released to the troposphere is deposited on the earth's surface (UNSCEAR, p. 24, 1962).

It has been found that many radionuclides with various half-lives are present in fallout debris. Those which have been designated as radiologically most important by the U.N. Scientific Committee on the Effects of Atomic Radiation (1962) are zirconium-95, niobium-95, and cesium-137. Of these, the external dose from cesium-137 may be of most concern because it decays mainly after deposition on the ground due to its long half-life.

Strontium-90, cesium-137, iodine-131 and carbon-14 are regarded as the main contributors to internal exposure (UNSCEAR, p. 25, 1962). In addition, strontium-89 has been added to the list of radionuclides considered as particularly important in the problem of radioactive contamination of food by fallout (FRC, 1965). The internal exposure problem from these isotopes is accentuated by their ability to undergo a biological concentration process in a variety of plants and animals which are important in the food-chain of humans.

Detailed information on fission product deposition and dietary levels of some radionuclides is available for a large city in the midwest, Chicago, which was one of the study areas in the Tri-City Sampling Program of the Health and Safety Laboratory of the AEC. Gustafson, Brar, and Muniak (1965) studied the ^{137}Cs content of quarterly food samples from Chicago supermarkets by means of gamma-ray spectrometry. A sudden increase in dietary intake during October 1961 followed the heavy fallout that occurred in late September 1961. In 1962 the highest dietary level was revealed in the July sampling and was in large part due to the concentration of ^{137}Cs in milk. The highest level was reached in November of

1963 and was attributed to a general elevation in all food items sampled. A lag between body burden and dietary level of ^{137}Cs was observed in measurements of adult Chicago residents in the whole-body counter of the Argonne National Laboratory. The lag was due to a biological half-life for ^{137}Cs in the human body of approximately 100 days in adults. Cesium is primarily distributed in the viscera and musculature because it is chemically similar to potassium. However, Anderson and Gustafson (1962) studied samples of human rib bone taken from various age groups ranging from infants to adults over sixty years. They found the highest concentrations of ^{137}Cs in the bones of the very young.

In summary, the Chicago study found milk to be the most important single dietary contributor of ^{137}Cs after the 1961 resumption of nuclear testing in the atmosphere. A fivefold increase in dietary level between 1961 and late 1963 was paralleled by an increase in whole-body burden with a time lag of six to nine months.

Cesium-137 levels in the food-stuffs produced in northern regions are generally higher than those expected from the amounts of the radionuclide deposited per unit area and have been found to be especially high in reindeer and caribou meat, as well as in fish from lakes low in mineral content (Kolehmainen, S., Häsänen, E. and Miettinen, J.K., 1967).

The question of the biological concentration of Cesium-137 in fishes from the high fallout area of northwestern Minnesota was investigated during the years 1954-1965 (Gustafson, 1969). This radionuclide has a relatively short retention time in humans (100 days in adults) but is of special radioecologic concern because as it traverses through the food chain there is an increase in the $^{137}Cs/K$ ratio through each trophic level (Pendleton and Hanson, 1958). Gustafson (1969) found relatively high levels (3,400-18,000 pCi/kg) of ^{137}Cs in bottom sediments from Red Lake in 1965. He also confirmed the trophic level biological magnification for 137 Cs in Red Lake's fish: mixed small fish had levels of 410 pCi/kg, perch had 750 pCi/kg, and the northern pike at the top of the aquatic food chain had a level of 3,620 pCi/kg. Considering that the level of ^{137}Cs in the Red Lake water was only about 1 pCi/kg, it is obvious that biological concentration and food-chain magnification are highly important in considering the sources of internal radiation exposure to Minnesota residents.

The isotope which so far has been of greatest significance in terms of short-term dose commitment to the population of the northern hemisphere is strontium-90. The global annual deposition of this radionuclide has been estimated from the results of two worldwide networks operated by the United Kingdom Atomic Energy Research Establishment (AERE) and the United States Atomic Energy Commission Health and Safety Laboratory (HASL). The latter network was found to have underestimated the deposition by as much as twenty per cent (20%) (see UNSCEAR, p. 15, 1969). Table 2 clearly reveals that 1963 was the peak fallout year with 2.62 megacuries of ^{90}Sr being deposited in the northern hemisphere. The worldwide cumulative deposit as of 1967 was 12.31 megacuries and 9.62 megacuries had accumulated in the northern hemisphere.

In the period 1965-1967, the annual deposition of ^{90}Sr decreased by about fifty per cent (50%) per year in the northern hemisphere. The latitudinal variation showed the highest levels of deposition in the middle latitudes of each hemisphere. A marked seasonal variation was observed with the maximum recurring in the spring.

During the period up to the middle of 1967, most of the deposition of ^{90}Sr and ^{137}Cs was due to tests occurring before 1963 (UNSCEAR, p. 15, 1969). Steadily increasing $^{144}Ce/^{137}Cs$ ratios in the northern hemisphere indicated that most of the long-lived radionuclides came from recent explosions. It was further estimated that approximately half of the 1968 deposition of long-lived radioisotopes came from recent explosions (Cambray, et al.).

Diet, including both animal and vegetable foods, is the major source of ^{90}Sr in human populations; inhalation and drinking water generally make only a minor contribution (UNSCEAR, p. 25, 1962). Biological concentration of this radionuclide in the flesh of fish (Krumholtz, 1956; Huver, 1970) is a potentially serious source of human dietary intake. After ^{90}Sr has been absorbed from the gastrointestinal tract, its distribution in the body is similar to that of

TABLE 2. Annual and cumulative world-wide ^{90}Sr deposition

(values in megacuries)

	Annual deposition			Cumulative deposit		
	North'n hemisph.	South'n hemisph.	Total	North'n hemisph.	South'n hemisph.	Total
Pre-1958				1.7	0.6	2.3
1958	0.74	0.31	1.05	2.39	0.83	3.22
1959	1.10	0.19	1.29	3.41	0.99	4.40
1960	0.26	0.17	0.43	3.59	1.14	4.73
1961	0.35	0.19	0.54	3.84	1.28	5.12
1962	1.45	0.31	1.76	5.16	1.55	6.71
1963	2.62	0.33	2.95	7.62	1.84	9.46
1964	1.66	0.44	2.10	9.06	2.21	11.27
1965	0.78	0.36	1.14	9.61	2.51	12:12
1966	0.33	0.21	0.54	9.70	2.65	12.35
1967	0.17	0.11	0.28	9.62	2.69	12.31

calcium so that it is deposited in bone and teeth. The concentration in new bone is related to the ratio of strontium-90 to calcium in the diet and some discrimination against ^{90}Sr absorption occurs in the intestine so that the observed ratio of strontium-90 to calcium in bone is only one-quarter of that found in the diet.

The ratio of strontium-90 to strontium-89 in worldwide fallout varies greatly depending on the location and magnitude of the weapons being tested, but a mean ratio of about 10 has been estimated for the years 1955-1960 (Crooks, et al., 1961). Thus strontium-90 has much exceeded strontium-89 in terms of population radiation dose. Although, it has been reported in certain instances that over short periods the dose rate from the shorter-lived nuclide was sometimes higher (Bryant, et al., 1958).

TABLE 3. ^{90}Sr inventory

	1963			1964		
	Mar.	July	Nov.	Mar.	July	Nov.
Stratosphere...........	6.5	5.1	3.8	3.0	2.1	1.7
Troposphere...........	0.5	0.2	0.2	0.4	0.3	0.1
Local fall-out..........	2.4	2.4	2.4	2.3	2.3	2.3
Global fall-out	7.2	8.7	9.4	9.9	10.9	11.2
Total	16.6	16.4	15.8	15.6	15.6	15.3
Corrected for/ decay to March 1963	16.6	16.5	16.1	16.0	16.1	15.9

If equal activities of strontium-89 and strontium-90 are present in bone the radiation dose rate from the latter is about twice as great because of the added beta radiation of its daughter, yttrium-90. This daughter product unlike its parent tends to concentrate in the soft tissues such as the gonads (Muller, 1967).

A radionuclide of serious concern in fresh fallout is iodine-131. This short-lived isotope (half-life 8.5 days) is of special interest because it concentrates in the thyroid gland. Iodine-131 is a potentially hazardous radionuclide because it undergoes a radioecological concentration process from plants through milk to the thyroid glands of humans.

The kinetics of environmental radioiodine transport through the milk-food chain has been studied experimentally by C.A. Pelletier and J.D. Zimbrick (1970). A major difference between adults and children was noted in the ingestion to inhalation thyroid dose ratio which was 47 for a standard man and 910 for a child in the 0-1 year range. The extreme sensitivity to continuous exposure to 2.6×10^{-13} μCi/cc of iodine-131 would result in an ingestion does of 1.5 rem/yr to the thyroid of a 0-1 year child.

It has been pointed out that estimates of mean population dose from this source are of little interest because considerably higher doses are delibered to infants than to adults (Loutit and Russell, 1966). The thyroid gland of an infant aged six months weighs about 2 g, while that of an adult weighs 25 g. Since the major portion of iodine binding will take place in the thyroid gland, the same intake of iodine-131 will deliver to the infant's thyroid more than ten times the dose received by adults. Because the young

TABLE 3. ^{90}Sr inventory

	1963			1964		
	Mar.	July	Nov.	Mar.	July	Nov.
Stratosphere	6.5	5.1	3.8	3.0	2.1	1.7
Troposphere	0.5	0.2	0.2	0.4	0.3	0.1
Local fall-out	2.4	2.4	2.4	2.3	2.3	2.3
Global fall-out	7.2	8.7	9.4	9.9	10.9	11.2
Total	16.6	16.4	15.8	15.6	15.6	15.3
Corrected for/ decay to March 1963	16.6	16.5	16.1	16.0	16.1	15.9

often consume greater volumes of fresh milk, the difference between infants and adults may be further increased.

The question of thyroid gland exposure in children first received wide attention after the major nuclear weapons tests in late 1961 when it was discovered in the United States and in the United Kingdom that children under the age of one year, who were fed on fresh milk, may have received about 170 mrad during a period of three months (UNSCEAR, 1962).

A radioisotope of considerable long-term concern is carbon-14. Because of its long half-life (5,760 years) the long-term dose commitment to the human population over a span of many centuries will greatly exceed that of other fallout components. In 1963, the carbon-14 content of the atmosphere was 90 percent above the natural background level, but it is predicted (UNSCEAR, 1964) that the corresponding value will be about 3 percent by the year A.D. 2000 because of equilibration with the abundant supply of non-radioactive carbon in the earth's oceans. J.F. Loutit and R.S. Russell (1966) have predicted that, "20,000 years will pass before 90 percent of the total dose from carbon-14 formed in past weapons tests have been received by the human race and that, summed over an infinite period, the total dose from this nuclide will exceed that which all fission products will have delivered by the end of the twentieth century."

Zinc-65 is an activation product which has been reported from nuclear reactor effluents and in worldwide weapons testing fallout. Perkins and Nielson (1959) found that zinc-65 in the Columbia River from effluents of the Hanford Works passed in measurable amounts to crops grown on land where the water was used for irrigation. The activity of zinc-65 per unit dry weight of pasture grasses reached more than 400 times that in water. The activities in milk and beef from animals grazed on this pasture were more than 25 times that in the water. Relatively small amounts of zinc-65 from weapons testing fallout has been reported in meat, milk and vegetables, but significant levels have been reported in marine foods (Murthy, et al., 1959).

A radionuclide which is only infrequently discussed in relation to fallout is tritium. This is surprising for tritium is the radioisotope which has been found in the highest concentrations in the surface waters of Minnesota by the monitoring programs of the U.S. Geological Survey and the Southwest Radiological Health Laboratory of the U.S. Public Health Service. Tompkins (1971) has reported that fission weapons release 1,000 to 2,000 curies of tritum per megaton, while fusion weapons release 6 to 10 megacuries per megaton. The worldwide inventory in 1960 was approximately 700 megacuries largely resulting from testing in the 1950 decade. The inventory in 1970 was estimated at about 2,000 megacuries (Tompkins, 1971). Considering that the total inventory at equilibrium of cosmic ray produced tritium is only 50 megacuries, it is apparent that the present relatively high degree of contamination of surface water (300 to 1,000 pCi/l) is due to weapons testing. At the height of weapons testing in 1963, tritium levels in the Mississippi River at Anoka reached a high of 6,240 pCi/l (Wyerman, et al., 1970) as compared to the 3.2 pCi/l estimated to have been present before nuclear testing in the atmosphere.

Tritium is also the predominant radioisotope released to the environment in the liquid waste discharges of light water reactors. The degree of contamination at a reactor site will largely depend upon the amount of diffusion through the fuel elements into the primary coolant (Jacobs, 1968). Such factors as cladding defects and the nature of the neutron-slowing poison added to the primary coolant are also important. Because tritium has a half-life of 12.26 years, contamination of the atmosphere and surface water with this radioisotope is a relatively long-term problem.

Tritium generally enters the body in the form of tritiated water (THO) and is transported through a variety of metabolic pathways to become widely distributed and incorporated into a wide array of biological molecules. Smith and Taylor (1969) surveyed the biochemical literature on metabolic pathways to find in utilizing tritiated water at least seven of the reactions during glycolysis would be expected to incorporate tritium into the carbon-hydrogen bonds of intermediary products. Many of these products would eventually participate in the biosynthesis of lipids, proteins, and one-carbon units. They further reported that in mammalian tissues, one-carbon unit metabolism would contribute tritium to intracellular lipid and nucleic acid constituents. It is becoming recognized that tritium incorporated into molecular species such as nucleic acids, may produce a much greater amount of injury to the cell than a more generalized distribution of equal amounts of energy from exogenous radiation.

Because tritium becomes incorporated in the DNA molecule (among others) the problem of genetic mutation and chromosome damage has to be faced. For instance, Gray (1959) reported the remarkable result that B-rays of the energy of tritium are about 2.5 times as effective in producing as are gamma rays. Plaut (1959) has attributed the high chromosome breakage ability of tritium as compared to ^{14}C to its higher ionization density (200/μ as compared to 30/μ).

That tritium can cause increased tumor formation in mice has been demonstrated by Lisco, et al. (1961) and Baserga, et al. (1962). Upon injection of a 1μCi/gm dose of ^{3}H-thymidine, they found that significantly more of these animals died from tumors than controls. Very low doses, as low as 0.3μCi/gm, of ^{3}H-thymidine, when injected at birth, significantly increased the incidence of lymphosarcomas in mice (Mewissen, 1970).

Doubtless, the most serious type of biological damage which has been demonstrated for tritium is that of genetic mutations. Huver (1971) has summarized the literature on this aspect of the tritium hazard. By means of injecting ^{3}H-thymidine into the testes of male mice, Bateman and Chandley (1962) estimated that 1% of the tritium disintegrations produced a dominant lethal mutation in the sperm.

It is clear that based upon the biological evidence that tritium is a classic paradigm of being a serious environmental contaminant whose significance has not been fully realized. It has unfortunately often been downplayed by those whose support of nuclear power plants and nuclear explosions overshadows their concern for public health.

Another source of environmental contamination and radiation exposure to Minnesota residents has been the Elk River reactor. Minnesota's introduction to the problems of nuclear reactor technology came with the construction of the Elk River reactor which was completed in

1963. The 22,000 kilowatt boiling water reactor was built by the Allis-Chalmers Manufacturing Company. The reactor is owned by the AEC but has been under the operating management of the Rural Cooperative Power Association. The reactor has been decommissioned and has been dismantled under the direction of the AEC, yet low-level radioactive discharges to the Mississippi River occurred from the contaminated remnants of the defunct reactor during the dismantling process.

The nuclear plant which began producing electricity during the summer of 1963 has been faced with a series of shut-downs. In October of 1963, the plant was down because of repairs in the primary heat exchanger. The reactor had reached about thirty percent of capacity when trouble involving trapped steam in water pipes was discovered. Cramer (1968) in a summary of safety-related occurrences in reactors noted two incidents with respect to the Elk River reactor. According to Docket 115-1, June 28, 1966, E.E. Wolters, twenty-six of the 100 core 1 elements showed some degree of bowing, with three showing bowing greater than 100 mils. The fuel tubes were 450 mils in diameter, with 300 mil spacing. The greatest displacement due to bowing was 150 mils. All twenty-six elements were replaced.

A second incident occurring in 1966 (Rural Cooperative Power Association, Elk River Monthly Operating Report No. 48, USAEC Report Coo-651-37, October, 1966) involved high coolant iodine levels. Coolant iodine levels indicated that some 9 g of ^{235}U was exposed due to cladding defects.

Radioactive leaks finally forced the closing of the Elk River reactor in December, 1968, after an intermittent operating history over a period of five years. During the operating history of this reactor, there were sharp increases in the amounts of radioactive wastes released to the environment. Gaseous radioactive wastes were released to the atmosphere and liquid radioactive effluents were released to the Mississippi River in a periodic batch disposal process. Solid wastes were taken off-site by the AEC to specified storage and disposal areas.

Data provided by the joint monitoring program of the Rural Cooperative Power Association and the Minnesota

Board of Health reveal marked increases in the amounts of radioactive wastes discharged to the air and to the Mississippi River as the operational life of the reactor increased. In terms of cumulative gallons of radioactive water discharged to the environment from near the end of 1963 through 1968, there has been a total release of 224 million gallons of waste water. More importantly, there was a rise in the radioactivity per unit volume of waste water discharged to the river.

The releases from the Elk River reactor have been relatively small when compared with the waste release histories of several other boiling water reactors such as Dresden-1, Big Rock Point, and Humboldt Bay, and the current gaseous releases of the Monticello reactor. The exposures from the Elk River releases must have been very small when compared with the exposures during the height of the weapons testing in the year 1963 or with diagnostic medical exposures which according to K.Z. Morgan (1962) deliver a genetically significant dose to the U.S. population of 55 mrem/yr. Nevertheless, there is no scientific evidence that the releases from the Elk River plant have been harmless.

The International Commission on Radiological Protection (See ICRP Publication 14, pp. 20-21) has taken the position that it is unlikely that the dose-responses for genetic effects and cancer have any sort of threshold. The former director of the Health Physics Division of the Oak Ridge National Laboratory, Dr. K.Z. Morgan (1968), agrees with the above ICRP position because, "present scientific evidence seems to indicate there is no threshold level of exposure to any form of ionizing radiation so low that the risk of radiation damage becomes zero. In other words, there are certain types of radiation-induced risks such as leukemia, bone tumors, thyroid cancers, and genetic damage that seem to relate more or less linearly with dose."

The question to be answered in this study is whether low-level radiation (levels below natural background, 130 mrem/yr or less) can influence the incidence of certain forms of cancer known to be radiation sensitive at higher dosages and whether such low dose exposures can influence the rates of malformations at birth and infant mortality in Minnesota human populations.

METHODS OF STUDY

A meaningful study of the geography of disease in relation to the environment requires an ecological approach. Unfortunately, the traditional subject of ecology has had too narrow a focus to adequately handle most relevant human environmental problems, therefore; an interdisciplinary approach is needed. R.W. Armstrong (1971) in an essay on "Medical geography and its geologic substrate," has emphasized this viewpoint:

> The relative importance of geochemical variables versus other environmental factors is still a guess. The only point of general agreement is that these chronic diseases have a complex, multifactorial etiology which makes it important to examine any particular set of variables in their ecological context. . . . Human ingestion of rock-derived elements, that may or may not influence the susceptibility or resistance of the body to disease, is influenced by a host of physical, biological, and cultural factors ranging from climate, through plant and animal metabolism to cooking practices. Further, there are the individual characteristics of man himself, particularly his physiology, his selectivity, and his capacity to adapt to environmental challenges. Time and space are significant in this system because events move at different rates and rhythms and have wide-ranging origins. A particular person's diet, for example, will usually originate from a number of localities both near to and far from his normal place of residence. To explore the intricacies of this ecological system would require the combined efforts of scientists drawn from several basic and applied fields, each employing coordinated research strategies which would allow subsequent integration of the separate studies to provide a better understanding of the whole system. An interdisciplinary approach seems the only feasible way of probing further into the question of geochemical implications with disease, of evaluating their relative importance to other environmental variables, and of proving or refuting the validity of the associations which appear suggestive of vital clues to the etiology of major chronic diseases."

The Federal Radiation Council (1960) has among its recommendations that to clarify the most critical problem areas concerning the effects of radiation on humans, special attention should be given to the following research approaches:

1. Increasing epidemiological studies on humans who have been exposed to radiation especially in doses sufficient to offer some probability that deleterious effects can be found.

2. Studies designed to determine more adequately the relationship between damage and dose at low total dose and low dose rates.

An interdisciplinary effort will be made in the northwestern Minnesota study combining the ecologic, epidemiologic, and biophysical approaches with the techniques of the social science field investigation. A broadly trained biologist with experience in genetics, embryology, ecology, and cancer research will be the project coordinator. A statistician experienced in the epidemiologic approach to studies of radiation effects will handle the statistical methodology of the study. A social scientist will be in charge of the field investigation of dietary habits in the contaminated region. A biophysicist will utilize the results obtained in the dietary survey, environmental radiation levels, and other pertinent information provided by the biologist-coordinator to establish the relationship between the doses received by the exposed population and certain forms of biological damage.

The forms of biological damage which will receive special attention in this study are:

1. Leukemia
2. Multiple myeloma
3. Cancer of the prostate
4. Cancer of the pancreas
5. Malformations at birth
6. Infant mortality

Leukemia, cancer of the prostate, and cancer of the pancreas were selected as forms of cancer likely to be responsive to changes in environmental radiation because of the major increases in the incidence of these diseases in Japan after the explosion of atomic bombs at Hiroshima and Hagasaki (see Figure 2 from Segi, et al., 1965).

FIGURE 2

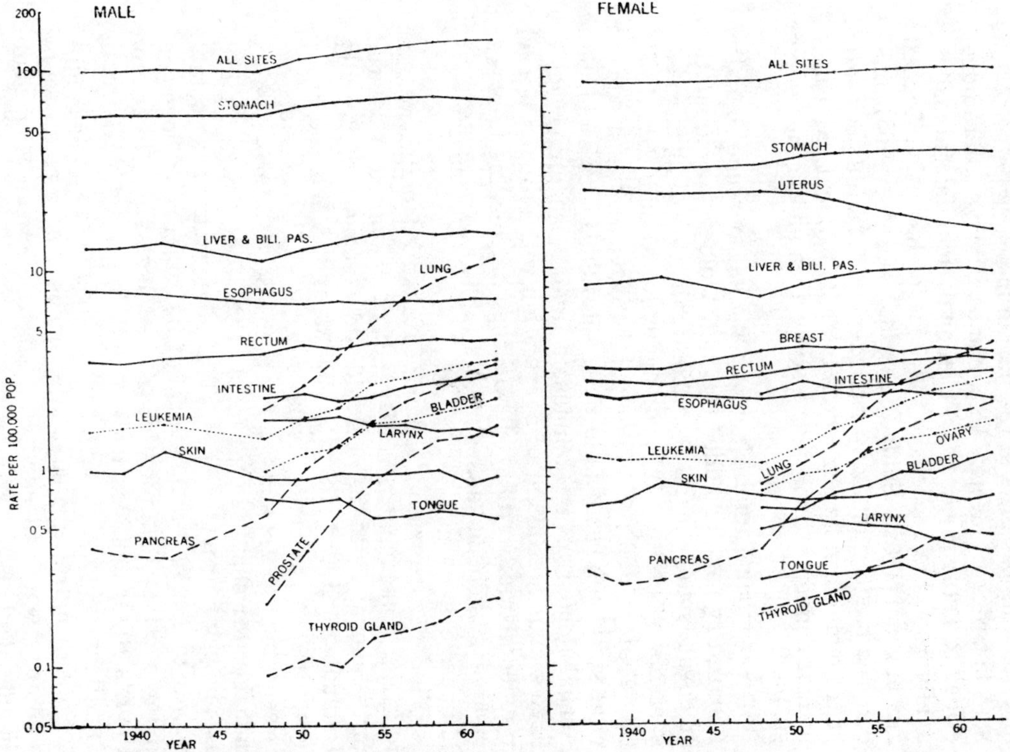

Trends in age-adjusted death rates for cancer of selected sites among Japanese (from Segi M., Kurihara M., and Matsuyama T. 1965)

The methodology to guide a field study of dietary intake of radionuclides of interest in fallout studies has been described by Thompson and Lengemann (1965). The primary purpose of the field survey would be to establish the eating and drinking patterns of the people in northwestern Minnesota in order to identify important pathways for radionuclides to enter the body. From analytical data already available in the reports of the Federal Radiation Council and in the Tri-City Study (Gustafson, et al., 1965), one can find basic radioisotope burdens of food for peak fallout years which can be corrected to apply to local environmental concentrations in the study area and in the control region.

Dr. Arthur Tamplin's group at the Lawrence Radiation Laboratory have provided the methodology for calculation of dosages to humans from contamination of aquatic ecosystems (Tamplin, 1969) or agricultural land (Ng, 1969; Ng, et al., 1968). If biological damages as revealed from vital statistics records in U.S. Vital Statistics or in the Annual Reports of the Minnesota Health Department do have a high coefficient of correlation with changes in environmental radioactivity, it will then be possible to make risk versus dosage assessments for the northwestern Minnesota situation.

Vital records on incidence of leukemia, malformations at birth, and infant mortality will be examined in relation to the annual releases of liquid and gaseous radioactivity (including tritium) from the Elk River and Monticello reactors. A convenient experimental situation exists between Minneapolis and St. Paul, for St. Paul has an additional water supply in the form of the Rice Creek Chain of Lakes which provides from 40 to 50 percent of the potable supply to St. Paul and several of its suburbs. Therefore, if radioactive contamination of the Mississippi River from the Elk River and Monticello plants has had a statistically significant impact on such health parameters as leukemia and infant mortality, such differences should be shown by a comparison of these vital statistics between Minneapolis and St. Paul.

A detailed multiple regression analysis between infant mortality and leukemia in comparison with age groups, sex, race, and income level will be attempted in addition to radioactive emission levels. The least squares method as modeled by DeGroot (1971) to correct for influence of fallout will be used in analyzing the possible effects of reactor effluents and emissions from Elk River and Monticello. For rural counties surrounding the Elk River or Monticello sites the following model is proposed:

$$M_t = B_0 + B_1 X_{1t} + B_2 X_{2t} + E_t$$

where t represents the particular year being studied during the span of operation of the reactor. Mt is the infant mortality rate in the county being studied for year t (i.e., the number of infant deaths per 1,000 live births in that year), and X_{2t} is a two year moving average for the years t and t-1 of the radioactive discharges measured in curies. X_{1t} is the infant mortality rate in Minnesota for the year t. Thus, we are regarding infant mortality (Mt) in a county near the Elk River or Monticello plants as a linear function of the infant mortality rate of the state of Minnesota plus a multiple of the discharges from the reactor.

It should be kept in mind that if a strong relationship between leukemia and infant mortality and reactor radioactive emissions could be established by this method, this would not constitute absolute proof of etiology. Such statistical evidence would, however, be influential in favoring the hypothesis that low-level emissions can affect leukemia and infant mortality rates.

Table 4. Environmental radioactivity sampling stations in Minnesota. (Minnesota Department of Health list)

SURFACE WATER ACTIVITY

Hallock	Fairmont	Crookston
East Grand Forks	St. Paul	International Falls
Eveleth		St. Cloud

MILK

Bemidji	Rochester	Duluth	Fergus Falls
Mankato	Worthington	Minneapolis	Little Falls

TRITIUM — Minnesota Department of Health

Bemidji — Upper Mississippi River
Hudson -- St. Croix River
Yucatan — Lower Mississippi
Steen — Des Moines and Missouri
Winnebago — Minnesota (Blue Earth)
Courtland — Minnesota River
Oslo — Red River
Harding — Lake of the Woods
Brookston — Lake Superior
Austin — Cedar River

TRITIUM — Southeastern Radiological Health Laboratory -- Terminated, January, 1971.

Clearwater — Mississippi River
Hallock — South Fork, Two Rivers
International Falls — Rainy River
Minneapolis — Mississippi River
Pine City -- St. Croix River 9 miles E
Red Wing — Mississippi River at Lock and Dam 3
Winona — Mississippi River 10 miles E
Fairmont — Budd Lake
Grand Forks — Red Lake River
Eveleth — St. Mary's Lake
St. Paul — Vadnais Lakes

TRITIUM — U. S. Geological Survey

Anoka — Mississippi River

LITERATURE CITED

Anderson, R.W. and Gustafson, P.F. 1962. Concentration of cesium-137 in human rib bone. Science, 137:668.

Armstrong, R.W. 1971. Medical geography and its geologic substrate. In Environmental Geochemistry in Health and Disease. H.L. Cannon and H.C. Hopps, eds. The Geological Society of America, Memoir 1923.

Barber, D.E. 1970. Natural differences in concentration of strontium-90 in milk. pp. 368-381. In Environmental Surveillance in the Vicinity of Nuclear Facilities. W.C. Reining, ed., Charles C. Thomas, Springfield.

Baserga, R., Lisco, H., and Kisieliski, W. 1962. Further observations on induction of tumors in mide with radioactive thymidine. Proc. Soc. Expt. Biol. Med., 110:587-590.

Bateman, A.J. and Chandley, A.C. 1962. Mutations induced in the mouse with tritiated thymidine. Nature, 193:705-706.

Bryant, F.J., Morgan, A. and Spicer, G.S. 1958. Radiostrontium in soil, herbage, animal bone and milk samples from the United Kingdom. U.K. Atomic Energy Authority Report, AERE-HPIR 2730.

Cambray, R.S., Fisher, E.M.R., Brooks, W.L., et al., Radioactive fallout in air and rain: results to the middle of 1968. Atomic Energy Research Establishment, Harwell report AERE-R 5899.

Cramer, E.N. 1968. Safety-related occurrences in reactor and radioactivity-handling operations as reported in 1966. Nuclear Safety, 9:249-265.

Crooks, R.N., Evett, T.W., Fisher, E.M.R., Lovett, M.B. and Osmond, R.G. 1961. Radioactive fallout in air and rain, results to the middle of 1961. U.K. Atomic Energy Authority Report, AERE-R 3766.

DeGroot, M.H. 1971. Statistical studies of the effect of low-level radiation from nuclear reactors on human health. Technical Report No. 54. Presented at the conference on "Planning an Epidemiological Study of Pollution Effects," a section of the Sixth Berkeley Symposium on Mathematical Statistics and Probability, Berkeley, California, July 19-22, 1971. 20 pp.

Drevinsky, P.J. and Pecci, J. 1965. Size and vertical distributions of stratospheric aerosols. pp. 158-182. In Radioactive Fallout from Nuclear Weapons Tests, A.W. Klement Jr., ed., AEC Symposium Series No. 5.

Federal Radiation Council, 1960. Background material for the development of radiation protection standards. Report No. 1, 39 pp.

Federal Radiation Council, 1964. Revised fallout estimates for 1964-1965 and verification of the 1963 predictions. Report No. 6, 29 pp.

Federal Radiation Council, 1965. Background material for the development of radiation protection guides for strontium-89, strontium-90 and cesium-137. Report No. 7, 44 pp.

Gofman, J.W. 1972. The case against nuclear power. Catalyst for Environmental Quality, 2(3):18-20.

Gray, L.H. 1954. Some characteristics of biological damage induced by ionizing radiations. Radiation Res., 1:189-213.

Gustafson, P.F. 1967. Cesium-137 in freshwater fish during 1954-1965. pp. 249-257. In Symposium on Radioecology. D.J. Nelson and F.C. Evans, eds., Conf-670503 (TID-4500), Ann Arbor.

Gustafson, P.F., Brar, S.S. and Muniak, S.E. 1965. Variation of fission products and natural radioactivity in surface air. pp. 497-506. In Radioactive Fallout from Nuclear Weapons Tests, A.W. Klement, Jr., ed., AEC Symposium Series No. 5.

Huver, C.W. 1970. Biological effects of radioactive effluents. pp. 41-45. In The Ecology Controversy, G. McCuel and D.C. Bender, eds., Greenhaven Press, Anoka.

Huver, C.W. 1971. Biological hazards of tritium. Report to the Illinois Pollution Control Board on behalf of the Illinois Attorney General's Office, 5 pp.

International Commission on Radiological Protection, 1969. Radiosensitivity and spatial distribution of dose. ICRP Publication 14, 118 pp.

Jacobs, D.G. 1968. Sources of Tritium and its Behavior upon Release to the Environment. AEC Critical Review Series, TID 24635.

Kolehmainen, S., Häsänen, E., Miettinen, J.K. 1967. ^{137}Cs in fish, plankton and plants in Finnish lakes during 1964-1965, pp. 913-919. In Radioecological Concentration Processes. B. Aberg, F.P. Hungate, eds., Pergamon Press, Oxford.

Krey, P.W. and Fried, R.E. 1965. Long-range fallout from Sedan and Small Boy shots. pp. 82-97. In Radioactive Fallout from Nuclear Weapons Tests, A.W. Klement, Jr., ed., AEC Symposium Series No. 5.

Krumholtz, L.A. 1956. Observations on the fish population of a lake contaminated by radioactive wastes. Bull. Amer. Mus. Nat. Hist., 110:277.

Lilienfeld, A.M., Pederson, E. and Dowd, J.E. 1967. Cancer Epidemiology: Methods of Study. The Johns Hopkins Press, Baltimore, 165 pp.

Lisco, H., Baserga, R., and Kisieliski, W.E. 1961. Induction of tumors in mice with tritiated thymidine. Nature, 192:571-572.

Loutit, J.F. and Russell, R.S. 1966. Radiation in man's environment. pp. 3-26. In Radioactivity and Human Diet, R.S. Russell, ed., Pergamon Press, New York.

Martell, A.E. 1964. Iodine-131 fallout from underground tests. Science, 143:126-129.

Mewissen, D.J. 1970. Tumorigenic effects of small amounts of radiation. pp. 413-424. In Medical Radionuclides: Radiation Dose and Effects, Cloutien, R.J., Edwards, C.L., Snyder, W.S. and Anderson, E.B., eds., AEC Symposium Series, No. 20.

Minnesota Department of Health, 1968. Survey of environmental radioactivity, January-June 1968. Publication No. COO-651-64.

Minnesota Department of Health, 1969. Survey of environmental radioactivity, January-June 1969. Publication No. COO-651-81.

Morgan, K.Z. 1968a. Ionizing radiation: benefits versus risks. Presented at Annual Meeting of the Health Physics Society, June 16-20, 1968, 14 pp.

Morgan, K.Z. 1968b. Common sources of radiation exposure. The American Engineer, July 1968, pp. 38-42.

Muller, W.A. 1967. Gonad dose in male mice after incorporation of strontium-90. Nature, 214:931-933.

Murthy, G.K., Goldin, A.S. and Campbell, J.E. 1959. Zinc-65 in foods. Science, 130:1255.

Ng, Y.C. 1969. Estimation of the maximum dose to man from the fallout of radionuclides on agricultural land. pp. 95-124. In Biological Implications of the Nuclear Age, AEC Symposium No. 16.

Ng, Y.C., Burton, C.A., Thompson, S.E., Tandy, R.K., Kretner, H.K., and Pratt, M.W. 1968. Prediction of the Maximum Dosage to Man From the Fallout of Nuclear Devices. IV. Handbook for Estimating the Maximum Internal Dose From Radionuclides Released to the Biosphere, UCRL-50163, Part IV, Biology and Medicine, UC-48, TID-4500, 124 pp.

Pendleton, R.C. and Hanson, W.C. 1958. Absorption of caesium-137 by components of an aquatic community. pp. 419-422. In Proceedings 2nd International Conference of Peaceful Uses of Atomic Energy, United Nations, Geneva, 1-13 September 1958, 18:419-422.

Perkins, R.W. and Nielson, J.M. 1959. Zinc-65 in foods and people. Science, 129:94.

Plaut, W. 1959. The effect of tritium on the interpretation of autoradiographic studies on chromosomes. In Laboratory Investigation, 8:286-295.

Segi, M., Kurihara, M., and Matsuyama, T. 1965. Cancer Mortality in Japan (1899-1962). Tohoku University School of Medicine, Sendai, Japan, 125 pp.

Smith, T.E. and Taylor, R.T. 1969. Incorporation of tritium from tritiated water into carbohydrates, lipids, and nucleic acids, Lawrence Radiation Laboratory, UCRL-50781, 23 pp.

Tamplin, A.R. 1969. Estimation of the maximum dose to man from the contamination of an aquatic ecosystem with radionuclides. pp. 75-94. In Biological Implications of the Nuclear Age, AEC Symposium, No. 16.

Thompson, J.C., Jr. and Lengemann, F.W. 1965. Dietary intake radionuclides: effect of consumption patterns and evaluation by use of integrating samples. pp. 877-894. In Radioactive Fallout from Nuclear Weapons Tests, A.W. Klement, Jr., ed., AEC Symposium Series No. 5.

Tompkins, P.C. 1971. Environmental radiation protection criteria and tritium. Presented at Tritium Symposium, Las Vegas, August 30-September 2, 1971, EPA and University of Nevada.

United Nations Scientific Committee on the Effects of Atomic Radiation, 1962. Seventeenth Session, Suppl. No. 16 (A/5216), United Nations, New York.

United Nations Scientific Committee on the Effects of Atomic Radiation, 1964. Nineteenth Session, Suppl. No. 14 (A/5814), United Nations, New York.

U.S. Geological Survey, 1969. Water load of uranium, radium, and gross beta activity at selected gaging stations water year 1960-61. Water Supply Paper No. 1535-0.

Wyerman, T.A., Farnsworth, R.K., and Stewart, G.L. 1970. Tritium in streams in the United States, 1961-1968. Radiological Health and Data Reports, September 1970, pp. 421-431.

Toward a Realistic Fission Dose Estimate: Methodology and Case Study

by

Land Educational Associates Foundation, Inc.
Research Team:

Gertrude A. Dixon, M. A.
Naomi Jacobson
George I. J. Dixon, Ph. D

Chief Consultant:

John W. Gofman, Ph. D., M. D.

A summary of the findings of this study, NUCLEAR WASTE: THE TIME BOMB IN OUR BONES, was published by Land Educational Associates Foundation, Inc., 3368 Oak Ave., Stevens Point, Wis. 54481 (2/17/79; 2nd printing 3/12/79-$1.25)

CONTENTS

Introduction

This study consists of the development and application of a methodology for estimation of the actual radiation dose from nuclear fission to a given population. The fact that the world population is accumulating a radiation dose from radionuclides created by the fission process has been well known for many years. The knowledge that such contamination could be extremely hazardous was, in part, responsible for the cessation of most aerial bomb testing in 1963. In more recent years studies of the expected deleterious health effects from a given level of radiation have been expanded and refined. (1) The scientific community is in agreement that any additional radiation exposure will result in some added health risk. (2)

Despite the known hazards, estimates of actual doses to specific populations are sparse. Reliance has been placed on average dose estimates for the whole population of a country, latitude or the world. These estimates are derived primarily from hypothetical models which assume uniform dispersion of total curies of radiation released by bomb testing and the nuclear fuel cycle to large population groups.

Since it is well known that radiation is not uniformly dispersed, these estimates can be meaningless for prediction of hazards to actual people. In fact, research conducted fifteen to twenty years ago by the Agricultural Research Council, Medical Research Council and the United Nations concluded that in some areas doses from Strontium-90 could be up to ten times higher than "average" doses. (3) In 1958, J. L. Kulp and associates at Columbia University's Lamont Geophysical Laboratory found a bone sample with a radiostrontium level seventy-five times the average. (4) Even at this early stage of investigation the diets of particular populations such as rural populations that derive their food from local sources, Indian tribes of the upper Amazon, Eskimos, vegetarians, were identified as possibly having from two to twelve times more Strontium-90 than the average eastern U.S. urban population. (5)

Despite this early knowledge of much higher-than-average radiation doses to particular populations, the research was dominated by a concern to identify the risk to large urban populations and exceptions were dismissed as representing "small numbers of people." (6) In this pre-1963 Test Ban Treaty era the emphasis on assessing the dangers of bomb-testing or nuclear war to the survival of the human race, not of individuals, is understandable. However, the tendency to dismiss the importance of potentially large radiation doses to small groups of people has persisted to the present day. Thus, the U.S. Atomic Energy Commission, assessing the impact of radiation inadvertently released from underground bomb-testing in Nevada, reports high doses to "8 sheepherders," "14 individuals," "19 people," 3 ranches, etc., but concludes that doses did not exceed the Radiation Protection Guide "for the general population." (7)

The same tendency to discount radiation impact on individuals is also evident in the U.S. Environmental Protection Agency's (EPA) assessment of the health impact of the release of four radionuclides from the normal operation of nuclear reactors. Despite the admission that populations near nuclear facilities, subject to the "first pass" of effluents, are "exposed to higher concentrations" of radioactivity, the radiation dose and expected resulting health effects to these "local" populations are not reported. Instead, concentrations of radioactivity were "converted to population dose and then to health effects." The primary concern was effects on the large populations along the eastern seaboard. (8)

On the other hand, environmental impact statements for particular nuclear facilities presume an "impact" radius (usually 50 miles) within which most, if not all, of the expected radiation dose will be confined. Doses to this local population from other fission sources (past, present or future) are assumed to be the same as global or national dose estimates. Facility doses are calculated on expected releases of radionuclides, not on measurement of concentrations in the environment.

Actual sampling and analysis of radiation pathways to humans is very limited in terms of areas sampled, frequency, types of pathways and types of radionuclides. It is difficult in the absence of thorough and continual monitoring to arrive at a realistic radiation dose estimate to a particular group of individuals. In fact, the director of the federal body responsible for assessing the health and safety of siting nuclear facilities has stated that "it would be impossible for routine monitoring programs to determine actual doses to real individuals." (9)

However, it is the ACTUAL radiation dose which is expected to produce health risks to REAL individuals.

In 1960 W. O. Caster observed that "So far, the 'averagers' have dominated the national committees entrusted with setting safety standards [which]. . . introduces a dangerous concept: That the individual is not important, that only the average exposure per million population needs to be considered. . . . In several cases averages have been used to obscure the wide range of biological variation, and to make doubtful situations appear safe." (10) Two decades later the "average population" concept still dominates the methodology of fission dose estimates and concommitant health risks.

A basic Realistic Fission Dose estimate could serve many functions which extant estimates derived from hypothetical models cannot. Among these are:

1. Prediction of health effects (i.e., cancers, leukemias, genetic impact) to a given population could be made with more certainty.

2. Existing models for prediction of radiation effects could be refined by examining health effects already occurring in a population for which a Realistic Fission Dose estimate has been ascertained.

3. Decisions concerning additional radiation doses to a given population from new nuclear facilities could be guided by a more precise knowledge of radiation doses already committed to that population.

Notes

1. For a recent review of changing risk estimates from radiation, see Karl Z. Morgan, "Cancer and Low Level Ionizing Radiation," Bulletin of Atomic Scientists, September 1978.

2. After the National Academy of Sciences concluded in 1972 that no threshold for added cancer risk from radiation existed, there has been agreement on this principle by federal agencies.

3. R. Scott Russell, ed., Radioactivity and the Human Diet (Oxford, New York: Pergamon Press, 1966), p. 21.

4. John M. Fowler, ed., Fallout: A Study of Superbombs, Strontium 90, and Survival (New York: Basic Books, Inc., 1960), p. 58.

5. J. Laurence Kulp and Arthur R. Schulert, Strontium-90 in Man and His Environment, Volume I: Summary (Final Report: NYO-9934 Contract AT (30-1)-1656, Division of Biology and Medicine, U.S. Atomic Energy Commission (Palisades, New York: Geochemical Laboratory, Lamont Geological Observatory, Columbia University, May 1962), pp. 224-25; 311-314.

6. Kulp and Schulert, op. cit., p. 225.

7. See Robert E. Allen, Division of Operational Safety, U.S. Atomic Energy Commission, Summary Information on Accidental Releases of Radioactive Effluent to the Atmosphere from Underground Nuclear Detonations Designed for Containment, August 5, 1963 – June 30, 1971 (Appendix A, WASH-1183, UC-41) (Washington, D.C.: U.S. Government Printing Office, June 1971).

8. U.S. Environmental Protection Agency, Office of Radiation Programs, Environmental Radiation Dose Commitment: An Application to the Nuclear Power Industry (Washington, D.C.: U.S. Government Printing Office, February 1974).

9. Statement of Roger J. Mattson, Director, Division of Siting, Health & Safeguards Standards, U.S. Nuclear Regulatory Commission, presented at the Environmental Protection Agency Hearing on Proposed 40 CFR Part 190, March 8, 1976 (with Attachments), Attachment A, pp. 52-53.

10. W.O. Caster, "From Bomb to Man," in John M. Fowler, op. cit., pp. 49-50.

Section A:

Methodology

If a fission radiation dose is to approach reality, it must begin with actual environmental measurement of the radionuclides produced by the fissioning process. Unfortunately very few of the two hundred or so radioactive poisons created by splitting the atom are actually monitored in the human food chain. However, it is possible to approach a realistic estimate by utilizing available monitoring records as a base.

Of all the radiation pathways to humans from nuclear fission, the one most frequently monitored is fluid milk and the radionuclides most consistently analyzed are Iodine-131, Strontium-90 and Cesium-137. This is because fresh milk will more quickly reflect changes in levels of contamination, it is distributed to large numbers of people, and relatively uncomplicated and inexpensive analytical methods of radiation detection have been developed. It does not mean that milk consumption is responsible for most of the radiation dose – on the contrary, fresh milk consumption constitutes only a small proportion of the total dose. However, it presents the best available factual base for the methodology.

These three radionuclides (Sr-90, Cs-137 and I-131), then, will provide a quantitative base for the methodology. From recorded concentrations in milk the total diet dose can be determined for these three radionuclides – designated as the R 3 Diet Dose. Additional fission doses from a few other radionuclides in the diet and external dose can be estimated in proportion to the R3 Diet Dose.

R^3 Diet Dose Commitment (R^3 =3-Radionuclides: Sr-90, Cs-137, I-131)

First the radiation dose from foods is calculated. State health departments should be able to provide the concrete data base of specific concentrations (picocuries per liter) of I-131, Sr-90 and Cs-137 in milk. The methodology, simply, is to determine the total picocuries of each radionuclide in milk for a given period; add an average consumption of cheese with concentration factors for a Total Dairy picocurie intake. To this is added the Balance of Diet derived from studies estimating the proportions of dairy intake to other foods for each radionuclide. The product is the total R3 Diet picocurie intake for each radionuclide.

The total number of picocuries, thus derived for a given period in a particular area, is used to determine the R3 Diet Dose Commitment in millirems for each radionuclide to body organs and total body by consulting the U.S. Nuclear Regulatory Guide 1.109 tables listing millirem doses per picocurie intake. (1) Doses may be calculated for the four age groups outlined in Regulatory Guide 1.109. Diet dose commitments may be calculated for any year for which the basic milk monitoring data is available. The R3 Diet Dose Commitments for each year may be summed for an Accumulated Dose Commitment. The total Accumulated R3 Diet Dose Commitment can be calculated for any age group born since the beginning of milk monitoring, by using respective Infant, Child, Teen or Adult dose tables from Regulatory Guide 1.109. Numerical values for the R3 Diet Dose Commitment are presented in Table I.

TABLE I

NUMERICAL VALUES – R^3 DIET DOSE COMMITMENT

Basic Factors:	Average annual picocuries per liter of Sr-90, Cs-137, and I-131 in milk (pCi/l/milk/yr)[a]
	Milk consumption liters per year (l/milk/yr)[b]
	Consumption ratio: Cheese/Milk (CR)[c]

Concentration factor: Cheese/Milk (CF)[d]
Picocuries Non-Dairy sources: Percent of
Total Diet picocuries (%N-D)[e]

Iodine-131

pCi/l/milk/year times l/milk/year equals Total pCi/milk/yr.
Total/pCi/milk/yr divided by 42 (CR) times 2.6 (CF) equals
Total pCi/Fresh Cheese/yr. – Eliminate for Infant.
Total/pCi/milk/yr plus Total/pCi/Fresh Cheese/yr equals
Total Dairy/pCi/yr.
%N-D equals 30% for Child, Teen, Adult; 10% for Infant.
Total Dairy/pCi/yr plus Total Non-Dairy/pCi/yr equals
Total Diet/pCi/yr/Iodine-131.

Cesium-137

pCi/l/milk/yr times l/milk/yr equals Total/pCi/milk/yr.
Total/pCi/milk/yr divided by 42 (CR) times 3.75 (CF) equals
Total pCi/Fresh Cheese/yr. – Eliminate for Infant.
Total/pCi/milk/yr divided by 17 (CR) times 1 (CF) equals
Total/pCi/Cheddar Cheese/yr. – Eliminate for Infant.
Total/pCi/milk/yr plus Total/pCi/Fresh Cheese/yr plus
Total/pCi/Cheddar Cheese/yr equals Total Dairy/pCi/yr.
%N-D equals 71% for Child, Teen, Adult; 23% for Infant.
Total Diary/pCi/yr plus Total Non-Dairy/pCi/yr equals
Total Diet/pCi/yr/Cesium-137.

Strontium-90

pCi/l/milk/yr times l/milk/yr equals Total/pCi/milk/yr.
Total/pCi/milk/yr divided by 42 (CR) times 2 (CF) equals
Total/pCi/Fresh Cheese/yr. – Eliminate for Infant.
Total/pCi/milk/yr divided by 17 (CR) times 6 (CF) equals
Total/pCi/Cheddar Cheese/yr. – Eliminate for Infant.
Total/pCi/milk/yr plus Total/pCi/Fresh Cheese/yr plus
Total/pCi/Cheddar Cheese/yr equals Total Dairy/pCi/yr.
%N-D equals 66% for Child, Teen, Adult; 22% for Infant.
Total Dairy/pCi/year plus Total Non-Dairy/pCi/yr equals
Total Diet pCi/yr/Strontium-90.

To Calculate R^3 DIET DOSE COMMITMENT (millirems): Apply Total Diet pCi/year for each radionuclide to Tables E-11 through E-14 (millirems per pCi ingestion) for Infant, Child, Teen or Adult in U.S. NRC Regulatory Guide 1.109 to derive millirem/year dose commitment for organs or total body.[f] Add millirem/year doses for each radionuclide to derive TOTAL R^3 DIET DOSE COMMITMENT/YEAR.

A LIFETIME/ACCUMULATED/R 3 DIET DOSE COMMITMENT may be derived for populations born in the first year for which basic milk monitoring data is available by applying respective Infant, Teen, Child or Adult Tables for each year and totaling yearly dose commitments. The total ACCUMULATED/R 3 DIET DOSE COMMITMENT for other populations will represent only the addition to commitments prior to the period for which basic milk monitoring data is available.

---

[a] Monitoring data may include minus (-) figures indicating less-than-background radionuclide concentrations. These are unrealistic reflecting problems with laboratory analysis, and should be interpreted as zero (0). Data may also include "less than" figures (i.e., <5) reflecting relative uncertainty for radionuclide concentrations below the "less than" figure. These are to be avoided for averaging purposes -- raw data from the laboratory will provide exact picocurie figures.

[b] Consumption rates for milk vary in NRC Regulatory Guide 1.109 from 110 liters per year for an "average individual adult" to 400 liters per year for the "maximum exposed Teen." The factors chosen will depend upon the age group and reasonably expected consumption for a particular population considering region and dietary habits.

[c] The ratios of cheese consumption to milk are from U.S. Department of Agriculture, Household Food Consumption Survey, 1965-66, Report No. 1, Table 28, p. 203.

[d] Concentration factors of Cheese to Milk are from R. John Garner (Environmental Protection Agency, Radiation Office), Transfer of Radioactive Materials from the Terrestrial Environment to Animals and Man (Cleveland, Ohio: The Chemical Rubber Co. Press, 1973), p. 23.

[e] Picocuries of Dairy to Non-Dairy ratio in Total Diet: For I-131 from Garner, op. cit., p. 35 (average of estimated range); for Cs-137 from Merril Eisenbud, Environmental Radioactivity, 2nd Ed., (New York: Academic Press, 1973), p. 381; for Sr-90 from U.S. Environmental Protection Agency, Radiological Quality of the Environment in the United States, 1977 (EPA 520/1-77-009), p. 116.

[f] U.S. Nuclear Regulatory Commission, Regulatory Guide 1.109, Revision 1, October 1977: Calculation of Annual Doses to Man from Routine Releases of Reactor Effluents for the Purpose of Evaluating Compliance with 10 CFR Part 50, Appendix I.

Total Internal Dose Commitment

The internal radiation dose from ingestion of I-131, Sr-90 and Cs-137 (the factors in the R3 Diet) constitutes only a fraction of Total Internal Fission Dose. The United Nations Scientific Committee on the Effects of Atomic Radiation (UNSCEAR) reports expected worldwide dose commitments from atomic weapons testing. (2) Sr-90 and Cs-137 are identified in UNSCEAR reports as the major contributors to internal dose but three other specific radionuclides are also singled out as significant internal contributors. These nuclides – H-3, C-14, and Fe-55 – are estimated to account for 16.2% of the total internal whole-body dose and 15.3% of the total internal dose to bone-lining cells. Specific monitoring data for these nuclides is not usually available but the proportion of the internal dose which they contribute to a specific population can be estimated in relation to the Sr-90 and Cs-137 in the Realistic R3 Diet Dose.

The Total Internal Dose Commitment, then, is the sum of doses due to Sr-90, Cs-137, and I-131 from the R3 Diet plus a Balance of Internal Dose from H-3, C-14, and Fe-55. (See Table II.)

TABLE II

NUMERICAL VALUES – TOTAL FISSION DOSE COMMITMENT

Dose to Whole Body

Internal Dose:

R3 Diet Dose Commitment from Cs-137 and Sr-90 = 83.8%
Additional Internal dose from H-3, C-14, Fe-55 = 16.2%

External Dose:

2.27 X Internal Cs-137 in R3 Diet Dose = Cs-137 External Dose

1.048 X Internal Sr-90 in R3 Diet Dose = External Short-Lived Dose

Internal Dose plus External Dose plus I-131 from R3 Diet Dose = TOTAL FISSION DOSE COMMITMENT TO WHOLE BODY

Dose to Bone

Internal Dose:

R3 Diet Dose Commitment from Cs-137 and Sr-90 = 84.7%
Additional doses from H-3, C-14 and Fe-55 = 15.3%

External Dose:

2.27 X Internal Cs-137 in R3 Diet Dose = Cs-137 External Dose

.76 X Internal Sr-90 in R3 Diet Dose = External Short-Lived Dose

Internal Dose plus External Dose plus I-131 from R3 Diet Dose equals TOTAL FISSION DOSE COMMITMENT TO BONE

Note: The Sr-90 dose to bone marrow in the UNSCEAR report has been taken as representative of the Sr-90 dose to total body indicated in NRC Regulatory Guide 1.109. Dose to bone-lining cells (UNSCEAR) has been taken as representative of bone dose (NRC, 1.109).

Doses indicated by UNSCEAR from Kr-85 and Pu-239 are small and have been omitted from this calculation.

External doses from I-131 to total body and bone have not been calculated.

Adjustments to Total Internal Dose

Additional internal fission doses are difficult to quantify and will vary for individuals within the population. Some internal dose pathways which have been omitted from the Total Internal Dose will apply to all populations and, therefore, constitute a general conservatism in this methodology. Among these are the following:

1. Doses expected from other radionuclides: At least 187 radionuclides are produced by nuclear fission. (3) Of these only six have been given specific consideration. A great many more have been identified in the food chain. Among the more significant are Sr-89, Cs-134, Cs-144, Pu-239, Pu-240, Co-58, Co-60, Zn-65, all of which are released along the nuclear fuel and nuclear weapon production cycle. (4) The U.S. EPA estimates that Pu-239, 240 contributed 1.5 pCi/year to the New York City diet in 1974. (5) Highest intakes were from milk, meat, bakery products, fresh fruits and vegetables. Nuclides which have

been identified in cow's milk in addition to I-131, Sr-90 and Cs-137 (in the R3 Diet) are Na-22, Ca-45, Rb-86, Sr-89, Y-91, Ru-106, Te-132, Ba-140, Ce-144, Pb-203, Po-210, Ra-224. (6)

There are a great number of uncertainties about the distribution of these and many other fission products in the environment and their pathways to humans as well as the health impact. In 1968 scientists at Lawrence Radiation Laboratory (Livermore, California) began an attempt to consider "each and every radionuclide on the chart of the nuclides. In this respect, our estimates may indicate that a particular radionuclide is a hazard for one of two reasons: either (1) it will be a hazard because of what we know about it, or (2) it will be a hazard because of what we don't know about it." (7) Arthur R. Tamplin reported in 1974 that an important radionuclide, Cesium-136, had been completely overlooked in dose estimates from atomic bomb testing fallout. He concluded that "Cs-136 was a major contributor to the gonad, bone and bone marrow dose" from French bomb testing in the Pacific region. (8)

All told, additional fission doses from the unknown quantities and effects of other radionuclides could be substantial.

2. Internal dose from inhalation of nuclides: NRC's Regulatory Guide 1.109 lists doses per picocurie due to inhalation which could be applied if concentrations in air were known. However, this information is not available. Most radionuclides will contribute some dose from this pathway. It is considered especially significant for the insoluble nuclides and alpha emitters. (9) For uranium mills, fuel fabrication and reprocessing plants "internal exposure from inhalation pathways accounted for 66% to 99% of total body population dose." (10) According to EPA these doses may be extremely high in certain areas — up to 14,000 millirems per year to bronchial epithelium. These doses due largely to radon daughter nuclides are categorized by EPA as Technologically Enhanced Natural Radioactivity (TENR). Since this very significant "enhancement" occurs because uranium ore is mined and processed for fissioning (to produce nuclear fuel or weapons), it must be considered in the Fission Dose.

3. Radiation in Drinking Water: Fission products in water intake have not been included in the R3 Diet. Water supplies can be contaminated by direct deposition or by leaching into ground water. (11) Such nuclides as Pu-239, 240; tritium, Sr-90, Ra-222 have been identified in drinking water analyses. Levels of contamination could differ by a wide margin. The difficulty of keeping nuclides out of drinking water is illustrated by the fact that the EPA's proposal in 1975 to limit Sr-90 to 2 picocuries per liter was changed in 1976 to allow 8 picocuries per liter. (12)

4. Additional consumption of Dairy Products: The R3 Dairy Diet includes only fluid milk and cheese which constitutes only 69% of total dairy consumption. (13) The category of "ice cream and other frozen dairy products" alone is 10.5% of total dairy consumption. With growing popularity of these products (i.e., "yogurt parlors") this percentage has probably increased. Another dairy product growing in popularity, fresh cheddar "curds," will add more radioactivity than the normal aged cheddar cheese since short-lived nuclides will not have decayed. Some of these additional doses from omitted dairy products will accrue to all populations and will vary widely.

In addition to the above major omissions from the calculations for internal dose to the general population, substantial other additions may need to be considered for particular situations. As Merril Eisenbud has observed: "One must always be on guard for some subtlety that may influence the radiation exposure of populations." (14) Among possibly important factors are the following:

Foods eaten fresh will contain more radioactivity than food purchased some distance from production. The R3 Diet data assumes radioactive decay in foods purchased. Thus, farm families, eating fresh vegetables, fruits, berries, milk, cheese will have higher doses than calculated in the R3 Diet. It has also been observed that "fresh fruit and vegetables from . . . local fruit or vegetable stands may also be a potentially important source of I-131"(8-day half-life). (15) Individuals who often consume foods without cooking or washing them (i.e., wild berries, normally so consumed) will have higher doses. More time spent outdoors (work or leisure) will result in higher doses from inhaled beta particles.

Diet dose intake can be affected considerably by the differences in radionuclide concentrations among foods. For instance in meats, mutton and pork tissue will contain respectively 2 and 6.5 times the Cs-137 in beef tissue. (16) The most significant radiation dose from meat is venison — up to 13 times that of beef. A study at the Nuclear Fuel Services reprocessing plant estimated a whole-body dose commitment to individuals from Cs-137 and Cs-134 (mostly in deer meat) of 257 mrems compared to 1.7 mrems for an individual who consumed very little venison. (17) Substitution in whole or part of venison for beef in the diet could probably raise the total-body dose from diet more than any other single factor. Per unit consumed goat milk will yield five times as much I-131, 17.5 times as much Sr-90, and 25 times as much Cesium as cow milk. (18) Doses in grains can also vary widely. Whole wheat products can contain almost twice the Sr-90 as milled wheat products. (19)

The influence of dietary habits on radiation dose was recognized in early scientific investigations of atomic bomb fallout. Thus, Kulp and Schulert, reviewing findings in 1962, note that: "Rice, like wheat, has one of the highest concentrations of Sr-90 per gram Calcium per unit of fallout among common foods. Therefore, the diets of Southeast Asia are equal to those of New York although the fallout is only one-quarter. In continental China it is expected that the average diet will be at least twice that of New York." (20) The diet in parts of South America, they concluded, would contain Sr-90 concentrations "two to eight times higher per unit of fallout than in the eastern United States. This is due to the difference in the composition of the diet." (21) In Puerto Rico investigators found Sr-90 concentrations in plantain and bananas to be twenty-four times higher than in liquid milk. (22) A third of the Sr-90 in the diets of low income Puerto Ricans in 1961 came from plantain and bananas, and more than one-fourth from beans.

Soil and weather conditions can influence radioactive deposit. retention or resuspension in soil, uptake of radiation by plants and animals. Thus, Kulp and Schulert conclude that "in tropical areas of high rainfall, certain (perhaps most) plants such as yucca and platano, show Sr-

90 concentrations per unit of fallout ten to forty times higher than the total diets of the North Temperate zone." (23)

R. Scott Russell reports that "excepting lichens, the highest concentrations of fission products in plant forms were found in species grown in shallow water and marshes." The average picocurie per gram in marsh moss over dryland willow leaves ranged a factor of 2 to 10. (24) Foods produced from marshes (such as cranberries, rice, or animals fed marsh hay) will contain higher concentrations. Radioactivity in lichens will affect primarily wild game. Soils poorer in calcium will produce crops and animal food relatively higher in Sr-90 concentrations. Past deposits of Cs-137 in light, sandy soils are more readily resuspended to enter the food chain. (25)

The degree to which diet differences can influence radiation doses is summarized by Garner:

> Local situations, again particularly dietary habits, of both animals and man, have to be considered when attempting to predict the consequences of environmental contamination. It is clear that in the past the tendency to generalize has resulted in failure to recognize that there are local populations which may be at considerably greater — or for that matter, lesser — risk than the general population . . . . An accurate estimate of the consequences of any specific contamination event can only be obtained by conducting a field study. (26)

The Total Internal Dose Commitment represents the minimum dose which can be empirically derived from available food monitoring. Until more foods are tested, with more frequency, this is the best minimum estimate. However, considering the conservative factors discussed, the R3 Diet dose cannot represent the Total Fission diet dose.

Total Fission Dose Commitment

Radiation doses are received from external sources in addition to internal sources of ingestion and inhalation. NRC Regulatory Guide 1.109 lists dose factors to skin and whole body for external radiation but since data concerning concentrations in air or contaminated ground are not available, a definite calculation is not possible. The Total Fission Dose Commitment is estimated in this methodology by determining the ratio of internal to external dose with the R3 Diet Dose taken as representative of internal dose.

The United Nations Scientific Committee on the Effects of Atomic Radiation (UNSCEAR) has periodically published estimates of the total dose commitment to date from bomb testing. The latest of these reports covers the period before 1971. Earlier reports were made for periods before 1968 and before 1962. (27) The External dose in these reports is attributed almost entirely to Cs-137 and short-lived nuclides. The Cs-137 Internal Dose to Cs-137 External Dose ratio has remained fairly constant for all report periods. The External dose from short-lived nuclides was derived by UNSCEAR from studies of deposits of Sr-90. The ratio of Internal Sr-90 dose to External short-lived nuclide dose is consistent in the UNSCEAR reports for northern and southern temperate zones as well as world. The dose commitment estimates from short-lived nuclides have been steadily increasing in progressive reports while Internal doses from Sr-90 have been decreasing.

Most of the fission dose commitment throughout the world is thought to be due to fallout from nuclear weapons testing (annual dose estimates of weapons fallout, 4 millirems; nuclear fuel cycle, .003 millirems). It is, therefore, considered appropriate in this methodology to assume that the External/Internal radiation dose concluded by UNSCEAR is applicable to a Realistic Total Fission Dose estimate.

Estimated doses for external radiation from fission sources other than bomb-testing show a very wide range of uncertainty. For instance, direct external radiation from nuclear reactors ("gamma shine") "have been observed to vary from undetectable levels to values of the order of 1 rem/year (1,000 millirems/year)" according to the NRC. (28) The NRC also estimates that "newer" BWR plants are estimated to have "dose rates of the order of 20 millirem/year/unit." (29) In this Draft Environmental Statement (June 1976) relating to the LaCrosse reactor no information is given concerning estimates of "gamma shine" from the LaCrosse (Wis.) reactor which has been operating since 1969. In the Final Environmental Statement relating to construction of a new PWR at Tyrone (Wis.), NRC estimates that "direct radiation from the facility" will "generally be less than 5 millirads/year." (30)

The "gamma shine" factor which could particularly affect persons living close by is not addressed in Appendix I, to Part 50 of Title 10 of the Code of Regulations "design objective" rules. (The State of Minnesota noted the fact in the Public Hearing.) (31) These "design objective" rules of Appendix I do not apply to the legal limit of radiation dose to the public from reactors. The legally allowed dose, in 10 CFR Part 20, is 500 millirems per year whole body dose. The Nuclear Regulatory Commission clearly denies that the "design objective" doses in any way constitute limits of radiation dose: "the design objective quantities may be thought of as goals . . . there will be occasions when the design objective quantities may be exceeded."(32) The Appendix I "goals" impose rules limiting reactor operation when whole-body doses exceed 3 millirems/year from liquid releases or 5 millirems/year from gaseous releases. However, the NRC does not expect to be able to detect radiation doses to the public at these dose "goals" by environmental monitoring techniques. When the Environmental Protection Agency proposed lowering the legal limit of radiation dose to the public from 500 millirems/year to 25 millirems/year, the NRC stated that "it will be impossible to demonstrate, using environmental monitoring, either compliance or non-compliance at the low level specified in the [proposed 25 millirem/year] standard." (33) An undetermined whole-body external dose from "gamma shine" (range: undetectable to 1,000 millirems/year) is not discussed in the "goals" outlined in Appendix I.

External exposure also occurs via the gaseous plume of reactors. Appendix I "goals" for dose per year in the air from the gaseous effluents is 10 millirads/year for gamma radiation and 20 millirads/year for beta radiation. External exposure from the gaseous plume at the Dresden (Ill.) reactor was determined to be the most important pathway. (34) Doses to the skin are primarily from the gaseous plume. The NRC first proposed a skin dose "goal" of 5 millirems/year and a total air exposure "goal" of 10 millirads/year. However, in the process of setting Appendix I dose limits, the NRC found it necessary to revise the skin dose "design objective" from 5 to 15 millirems/year. (Skin dose estimates for Cases studied ranged up to 200 millirems/year at a distance of 1.24 miles from reactors.) (35) Appendix I "design objectives" now estimate no more than a total of 5 millirems/year total body dose from the gaseous plume, but the "design objective" potential total body exposure in the air is 30 millirads/year (10 gamma and 20 beta). Although gamma radiation is the greater hazard from external pathways, beta radiation, too, "can be a health hazard either as internal or external radiation." (36)

The range of possible doses to thyroid from the gaseous plume was very wide among Cases studied and the percentage of dose due to external source (including inhalation) ranged from practically none to 85% of the thyroid dose.

The range of possible dose to the public from noble gas releases from nuclear reactors is very wide. In 1977 the Environmental Protection Agency reported "site boundary doses" from operating reactors ranging from <1 to 77 millirems/year. (37)

Research concerning distances which may be affected by external radiation from nuclear facilities is scarce. If the nuclides are removed principally by rainout, as in the case of SO2 from coal plants, the radiation doses could extend for thousands of miles. Fallout from bomb-testing is expected worldwide. The transport and deposition of reactor fallout is not well researched. However, doses can be expected to be higher in downwind sectors and in rainout areas with doses concentrated closer to reactors in times of temperate inversions.

The UNSCEAR ratios of internal to external radiation dose seem to provide the best methodology for estimating External Dose, particularly in the light of a large range of uncertainty about External Dose estimates from nuclear facilities and the lack of thorough empirical studies. UNSCEAR calculations, then, will be used to determine what proportion of the Total Fission dose is represented by the R3 Diet Dose. External Cs-137 will be determined by proportion to Internal Cs-137 in the R3 Diet; External short-lived nuclide dose will be determined by proportion to Internal Sr-90 in the R3 Diet. The Total Fission Dose can be estimated for whole body or bone. Table II outlines numerical values.

Annual Internal Dose

The radiation doses discussed above and derived from dose tables in NRC Regulatory Guide 1.109 are 50-year Dose Commitment figures — the dose within 50 years after the ingestion of nuclides. To determine the part of the 50-year Dose Commitment which is received in a particular year the biologically Effective Half-life of each ingested nuclide must be considered. The term, Annual Dose, refers only to that part of the dose received (or decayed within the body) during a specific year. The Effective Half-life takes into account the combination of biological loss of the nuclide from the body plus the loss by radioactive decay. It can differ for each radionuclide, for route of entry, for solubility, for part of the body affected. NUREG-0172 provides tables with numerical values for all these aspects of Effective Half-life. (38)

For some nuclides all of the 50-year Dose Commitment may be received by the body within a few days — for others the dose may continue to be received every year for the whole 50-year period. For instance, practically all ingested I-131 dose commitment will be received within the year of ingestion; for Cs-137 approximately 80% will be received the first year with the remaining 20% carried to the second year after ingestion: for Sr-90 only approximately 4.4% will be received the first year after ingestion with the remaining 95.6% spread over a dose-commitment period of 50 years.

Table III and Figure I outline numerical values for deducing Annual Dose Received each year from a 50-year Dose Commitment. (39) Thus the R3 Diet Dose Commitment can be converted into Annual doses received in each year, with accuracy. The Annual Dose Accumulation for all years analyzed can then be summed for total-body or any organ. For epidemeological studies the actual Annual doses received will provide a useful data base. The Annual dose calculation is also more useful for purposes of comparison (i.e., to other estimates of Fission dose or "natural" radiation dose) since most radiation dose estimates are expressed in terms of Annual dose, not 50-year Dose Commitments.

TABLE III

CONVERSION OF Sr-90 50-YEAR DOSE COMMITMENT
BASE: EFFECTIVE HALF-LIFE (T-EFF) Sr-90 TO BONE:
6665 DAYS (18.26 YEARS)*

Year	Activity Left	% Dose Delivered	Cumulative %
1	1.0	4.4	4.4
2	0.97	4.3	8.7
3	0.93	4.1	12.8
4	0.90	4.0	16.8
5	0.86	3.8	20.6
6	0.83	3.7	24.3
7	0.80	3.5	27.8
8	0.77	3.4	31.2
9	0.74	3.3	34.5
10	0.71	3.1	37.6
11	0.68	3.0	40.6
12	0.66	2.9	43.5
13	0.63	2.8	46.3
14	0.61	2.7	49.0
15	0.58	2.6	51.6
16	0.56	2.5	54.1
17	0.54	2.4	56.5
18	0.52	2.3	58.8
19	0.50	2.2	61.0
20	0.49	2.2	63.2
21	0.47	2.1	65.3
22	0.45	2.0	67.3
23	0.43	1.9	69.2
24	0.42	1.8	71.0
25	0.40	1.8	72.8
26	0.38	1.7	74.5
27	0.37	1.6	76.1
28	0.36	1.6	77.7
29	0.35	1.5	79.2
30	0.33	1.5	80.7
31	0.31	1.4	82.1
32	0.30	1.3	83.4
33	0.29	1.3	84.7

Year	Activity Left	% Dose Delivered	Cumulative %
34	0.28	1.2	85.9
35	0.27	1.2	87.1
36	0.26	1.1	88.2
37	0.25	1.1	89.3
38	0.24	1.1	90.4
39	0.23	1.0	91.4
40	0.23	1.0	92.4
41	0.22	1.0	93.4
42	0.21	0.9	94.3
43	0.20	0.9	95.2
44	0.19	0.84	96.04
45	0.18	0.8	96.84
46	0.18	0.8	97.64
47	0.18	0.8	98.44
48	0.17	0.75	99.19
49	0.16	0.7	99.87
50	0.16	0.7	100.59

* T-EFF for Strontium-90 Dose Commitment to Whole Body is 5834 Days, slightly less than for Bone Dose. As a factor of conservatism, the same T-EFF outlined above for Bone Dose was als used for Whole Body Dose calculations in this study.

Calculations provided by John W. Gofman and L. Douglas DeNike based on NUREG-0172: G.R. Hoenes and J.K. Soldat, Age-Specific Radiation Dose Commitment Factors for a One-year Chronic Intake. November 1977. Table B-5.

The Annual Total Internal Fission Dose may be calculated by adding the ratio from ingested nuclides not included in the R3 Diet as outlined under "Internal Dose" for Whole Body or Bone in Table II.

Annual Total Fission Dose

A 1972 study by A.W. Klement, Jr., et al. (U.S. EPA) provides a methodology for computing Annual Total Fission Doses from the Annual R3 Diet Dose base. Ratios of Internal to External plus Inhalation for fallout doses are given for 1963, 1970, 1980, 1990, and 2000. For the period under study in Wisconsin (the years 1963-1976) the appli-

FIGURE I

STRONTIUM-90: EFFECTIVE HALF-LIFE 6665 days (18.26 years)

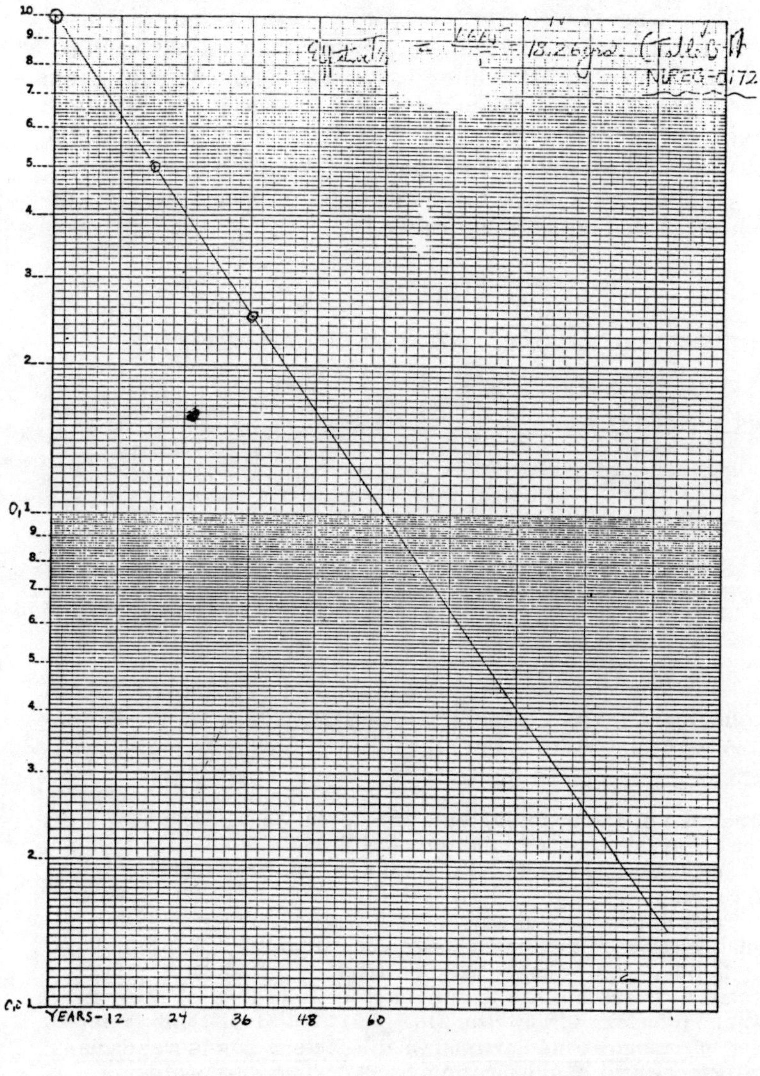

— Prepared by Dr. John W. Gofman

cable ratio of Internal to External dose is .917 to 1. For statistical convenience and conservatism a ratio of .9 to 1 is used to derive an Accumulated Annual Total Fission Dose for Wisconsin. For an Annual Total Fission Dose in any one year the appropriate ratio may be calculated.

No Internal to External ratios are indicated by Klement for annual radiation doses other than fallout in this study. However, all "Other" sources of non-natural radiation dose constitute less than 1% of the total non-natural dose. Therefore, application of fallout dose ratios to all sources in dose estimates should yield a good estimate.

Numerical Values for estimation of Annual Total Fission Dose are outlined in Table IV.

TABLE IV

NUMERICAL VALUES – TOTAL ANNUAL FISSION DOSE

Dose to Whole Body

Internal Dose: (1963-1976 or any one year)

R3 Diet Dose (Annual) from Cs-137, Sr-90	=	83.8%
Additional Annual Internal Dose	=	16.2%

External Dose: (1963-1976) Accumulation. *

Internal Dose plus I-131 from R3 Diet X .9 = External Annual Dose

Internal Dose plus External Dose = TOTAL ANNUAL DOSE ACCUMULATION

* The ratio of 1 to .9 for Internal to External Annual Dose is a value for average ratio for the years 1963-1976.

For applicable numerical value for any one year consult reference:

Table of Summary of Estimates of Whole-Body Environmental Radiation Doses to the United States Population in Klement, A.W., Jr., Miller, C.R., Minx, R.F., Schleien, B., Estimates of Ionizing Radiation Doses in the United States 1960-2000, U.S. Environmental Protection Agency Rep. ORP/CSD 72-1 (1972). Also referenced in Biological and Environmental Effects of Low-Level Radiation, Proceedings of a Symposium on Biological Effects of Low-

Level Radiation Pertinent to Protection of Man and His Environment, Organized by the International Atomic Energy Agency and the World Health Organization, held in Chicago, 3-7 November 1975. Volume II, p. 293.

Notes

1. U.S. Nuclear Regulatory Commission, Office of Standards Development, Regulatory Guide 1.109 (Revision 1, October 1977): Calculation of Annual Doses to Man from Routine Releases of Reactor Effluents for the Purpose of Evaluating Compliance with 10 CFR Part 50, Appendix I.

2. United Nations Scientific Committee on the Effects of Atomic Radiation. Ionizing Radiation Levels and Effects Volume I: Levels, (New York: United Nations, 1972) as reported in U.S. Environmental Protection Agency, Office of Radiation Programs, Radiological Quality of the Environment in the United States, 1977 (U.S. Environmental Protection Agengy:Washington, D.C.: September 1977). (EPA 520/1-77-009), p. 130.

3. G.R. Hoenes and J.K. Soldat, Age-Specific Radiation Dose Commitment Factors for a One-year Chronic Intake (NUREG-0172) (Battelle Pacific Northwest Laboratories, Richland, Washington: November 1977). Prepared for the Office of Standards Development, U.S. Nuclear Regulatory Commission, Under Contract No. B21446, p. B-1.

4. See R. John Garner, Environmental Protection Agency, Radiation Office, Transfer of Radioactive Materials from the Terrestrial Environment to Animals and Man (Cleveland, Ohio: The Chemical Rubber Co., 1973), for additional information.

5. U.S. Environmental Protection Agency, Office of Radiation Programs, Radiological Quality of the Environment in the United States, 1977 (U.S. Environmental Protection Agency: Washington, D.C.: September 1977)(EPA 520/1-77-009), (hereafter referred to as EPA, 1977), Table 4-6.

6. Garner, op. cit., p. 22.

7. Yook C. Ng, Burton, Thompson, Tandy, Kretner and Pratt, Prediction of the Maximum Dosage to Man from the Fallout of Nuclear Devices: IV. Handbook for Estimating the Maximum Internal Dose from Radionuclides Released to the Biosphere. (Livermore, California: Lawrence Radiation Laboratory, May 14, 1968), (TID-4500, UC-48, Biology and Medicine), p. vi.

8. Arthur R. Tamplin, "Random Violence in the Pacific," *Bulletin of the Atomic Scientists*, March 1974, p. 14.

9. Garner, *op. cit.*, p. 15.

10. *EPA 1977, op. cit.*, p. 145.

11. Garner, *op. cit.*, p. 15.

12. See EPA Drinking Water Standards as outlined in "Proposed Rules: National Interim Primary Drinking Water Regulations," *Federal Register*, p. 34327, August 14, 1975; and EPA, Office of Water Supply, *National Interim Primary Drinking Water Regulations* (Washington, D.C.: U.S. Government Printing Office, September 1976) (EPA-570/9-76-003).

13. U.S. Environmental Protection Agency, *EPA Assessment of Fallout in the United States from Atmospheric Nuclear Testing on September 26 and November 17, 1976 by the People's Republic of China*, August 1977. (EPA-520/5-77-002) Table C-3.

14. Merril Eisenbud, *Environmental Radioactivity, 2nd Ed.* (New York: Academic Press, 1973), p. 434.

15. Garner, *op, cit.*, p. 35.

16. Garner, *op. cit.*, p. 28.

17. *EPA, 1977, op cit.*, p. 178.

18. Garner, *op. cit.*, p. 22.

19. Garner, *op. cit.*, p. 38.

20. Kulp, *op. cit.*, p. 341.

21. Kulp, *op. cit.*, p. 220.

22. Kulp, *op. cit.*, p. 147.

23. Kulp, *op. cit.*, p. 220.

24. Russell, *op. cit.*, p. 402.

25. Garner, *op. cit.*, p. 40.

26. Garner, *op. cit.*, pp. 44-45.

27. Most recent UNSCEAR reports are summarized in EPA, Radiological Quality of the Environment in the United States, 1977 (see Note 2 above). The report covering the period prior to 1962 is summarized in R. Scott Russell, op. cit., p. 20.

28. U.S. Nuclear Regulatory Commission, Draft Environmental Statement Related to Operation of the LaCrosse Boiling Water Reactor (NUREG-0226 — Docket No. 50-409), June 1976.

29. Ibid.

30. U.S. Nuclear Regulatory Commission, Final Environmental Statement Related to Construction of Tyrone Energy Park (Docket No. STN 50-484), April 1977.

31. U.S. Atomic Energy Commission, Concluding Statement of Position of the Regulatory Staff — Public Rulemaking Hearing on: Numerical Guides for Design Objectives and Limiting Conditions for Operation to Meet the Criterion "As Low As Practicable" for Radioactive Material in Light-Water-Cooled Nuclear Power Reactors (Docket No. RM-50-2), February 20, 1974, p. 67.

32. From Statement by Roger J. Mattson, Director, Division of Siting Health & Safeguards Standards, U.S. Nuclear Regulatory Commission, presented at the U.S. Environmental Protection Agency Hearing on Proposed 40 CRF Part 190, March 8, 1976, Attachment A, p. 5.

33. Mattson, op. cit., p. 7. Mr. Mattson states further that "substantially increased environmental monitoring would be required and the economic risk of forced shutdown would be present throughout the operating history of a facility," op. cit., p. 6. It is unclear how these problems of radiation dose detection will be resolved if the new EPA dose limits are imposed as planned. For a further discussion see Gertrude A. Dixon, Comments on Draft Environmental Statement Related to Construction of Tyrone Energy Park (Stevens Point, Wisconsin: League Against Nuclear Dangers, August 14, 1976) or pages A-33 through A-44 in U.S. Nuclear Regulatory Commission, Office of Nuclear Reactor Regulation , Final Environmental Statement Related to Construction of Tyrone Energy Park (NUREG-0226), April 1977.

34. U.S. Department of Health, Education and Welfare, *Radiological Surveillance Studies at a Boiling Water Nuclear Power Reactor*, December 1969.

35. See U.S. Nuclear Regulatory Commission, WASH-1258, Tables 7-11, 7-12 and discussion in RM-50-2, *op. cit*.

36. U.S. Department of Agriculture, *Radiological Training Manual*, p. 25.

37. EPA, 1977, *op. cit.*, pp. 162-63.

38. Hoenes, *op. cit.*, Table B-5.

39. For this explanation of Annual Dose vs. Dose Commitment we acknowledge assistance from Dr. John Gofman. Dr. Gofman and Dr. L. Douglas DeNike contributed materials presented in Table III.

Section B:

Wisconsin Case Study

Study Design

The basic milk monitoring data necessary for estimation of a Realistic Fission Dose for Wisconsin has been available since 1963, when the Radiation Protection Section of the Department of Health and Social Services initiated milk monitoring for Iodine-131, Cesium-137 and Strontium-90 at six locations in the state. This study assumes that the major milk-producing areas of the state are represented by the distribution of the six stations and, therefore, the monitoring results should constitute a fairly correct estimate of concentrations of radionuclides in milk actually consumed by individuals.

Monitoring continued at five of the six stations through 1976. The LaCrosse station was dropped July 1, 1972. However, the Radiation Protection Section also monitors milk at the LaCrosse Boiling Water Reactor site (at Genoa, Wisconsin, 18 miles south of LaCrosse). Results of the reactor site monitoring were reported in Annual Environmental Survey Reports for the period May 1965 through December 1976, including milk monitoring at LaCrosse as a "Reference Station" from May 1965 through June 1971, and at "Genoa" (three dairy farms in the area) from July 1969 through December 1976. (1) For purposes of this study, it was decided that the best estimate of radionuclide concentrations in milk for the LaCrosse area would result from averaging all of the different concentrations reported for LaCrosse and Genoa.

FIGURE II

MONITORING STATIONS (MILK SHEDS) AND NUCLEAR FACILITIES

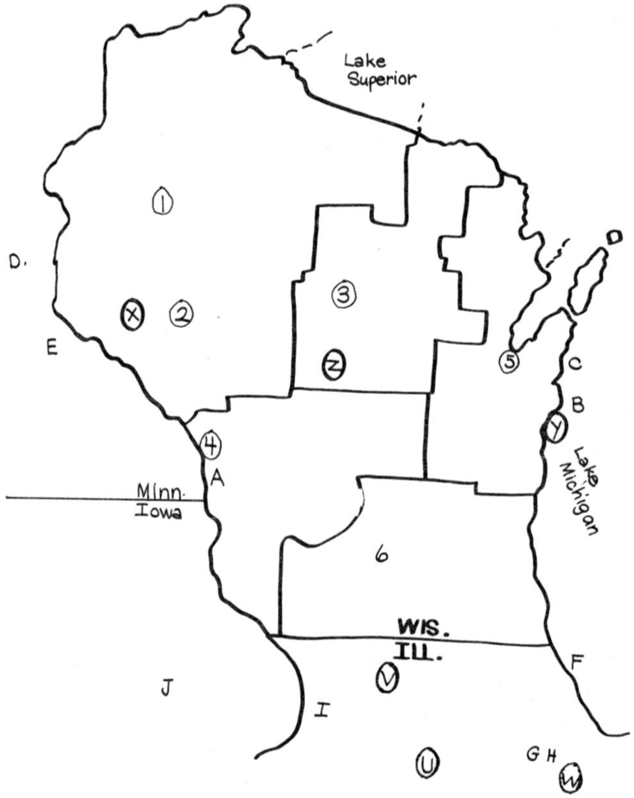

Milk Sheds
1 - Rice Lake
2 - Eau Claire
3 - Wausau
4 - LaCrosse
5 - Green Bay
6 - Madison
(Boundaries by probable County lines)

Nuclear Plants Operating
A - Genoa, Wis.
B - Point Beach, Wis.
C - Kewaunee, Wis.
D - Monticello, Minn.
E - Prairie Island, Minn.
F - Zion, Ill.
G - Dresden, Ill.
H - Morris, Ill.
I - Quad Cities, Cordova, Ill.
J - Arnold, Cedar Rapids, Ia.

Under Construction or Proposed
U - LaSalle (Seneca), Ill.
V - Byron, Ill.
W - Braidwood, Ill.
X - Tyrone, Wis.
Y - Haven, Wis.
Z - Rudolph, Wis.

The six stations are located on Figure II.

Since the data base for the R3 Diet Dose methodology is thus available in Wisconsin for a continuous period of fourteen years (1963-1976), the most comprehensive and realistic estimate for total accumulated fission dose could be computed for Wisconsin residents born in 1963. For residents born before 1963, a total lifetime accumulation cannot be as accurately estimated since no state monitoring records are available for doses committed in the years of nuclear fallout in the state before 1963. (2) However, for contrast, a separate calculation was planned for adult residents to estimate the dose added to pre-1963 radiation dose commitments.

Age-specific dose tables in Nuclear Regulatory Commission (NRC) Regulatory Guide 1.109 were used to calculate doses to Infant (1963); Child (1964-1973); Teen (1974-1976) for a realistic total dose to a 15-year-old during his or her lifetime. Another estimate using Adult dose (1963-1976) was made for 14 years of dose commitment to adult residents in addition to the unknown previous commitment. A milk consumption of 365 liters per year was assumed for Infant, Child, and Teen — 310 liters per year for Adult. (3)

Factors Influencing Realism of Wisconsin Fission Dose Estimates

The degree to which the R3 Diet Dose methodology can approximate reality is dependent on the degree to which the basic milk monitoring data represents actual levels of radioactivily ingested by real individuals. Therefore, monitoring of radionuclide concentrations in milk and local and regional exposure patterns were evaluated for possible significant influence on the derived doses.

Monitoring of Iodine-131, Cesium-137 and Strontium-90

Iodine-131. Milk was tested weekly through 1972 and bi-monthly thereafter with occasional exceptions of from only one test monthly up to eight. Frequency of testing should give fairly real concentration fluctuations for the non-nuclear stations. (4) However, the gamma-scan detection method used by the laboratory at these stations may yield dependably precise figures only at much higher concentrations of I-131 than are usually found in samples. (A superior chemical-extration method, used only at nuclear plant stations, is considered to be much more reliable but up to 80% of the I-131 present can be "lost" if **Formalin** is added to samples. See Note 5.)

Table V summarizes fluctuations and contradictions in reports by the laboratory concerning minimum detectable levels from 0 to 10 picocuries per liter. Computer printouts of milk monitoring at non-nuclear stations report all radionuclide concentrations in exact picocuries with no indication of a specific detection limit and no "less than" numbers. However, a report prepared for the League Against Nuclear Dangers (LAND) by the Radiation Protection Section (March 3, 1975) does not list exact picocurie amounts of weekly or bi-weekly testing but, rather, lists monthly averages indicating a detection limit of 5 picocuries per liter — all numbers below 5 pCi/l are indicated as <5 pCi/l. (6)

In the state Radiation Protection Section monitoring reports for nuclear plant stations, however, methods of reporting are quite different. From May 1965 through June 1972 radionuclide concentrations are given in exact picocurie amounts; July 1972 through December 1973 all concentrations below 5 pCi/l are indicated as <5 pCi/l; January 1974 through December 1976 all concentrations below 10pCi/l are indicated as <10p Ci/l. (7) The declared level of detection also changes from 3.2 pCi/l to 5 pCi/l to 10 pCi/l in these Annual Environmental Survey Reports by the state laboratory for nuclear plant stations.

TABLE V

VARIOUS STATED DETECTION LIMITS, CONCENTRATIONS REPORTED AND ANNUAL AVERAGING METHODS FOR I-131 IN MILK BY WISCONSIN RADIATION PROTECTION SECTION

All references to "gamma scan" analytical technique — "chemical separation" technique detection limits in Nuclear Plant Reports for 1975 and 1976 listed as 1 pCi/l; reported concentrations range from <0.8 to 5 pCi/l.
Gamma scan detection limit changes also apply to Cs-137 analyses.

Annual Environmental Radiation
Survey Reports for Nuclear Plants Sites

Date	Detection Level (pCi/l)	Lowest pCi/l Recorded	Averaging Method
5/1965 thru 6/1967	3.2	0	None
7/1967 thru 6/1969	3.2	<3.2	None
7/1969 thru 6/1972	3.2	0	Exact Numbers Averaged
7/1972 thru 6/1973	5	<5	<5
7/1973 thru 12/1973	5	<5	None
1/1974 thru 12/1976	10	<10	None

Computer Printouts — Non-Nuclear Sites

1963 thru 1976	None Stated	0	None Stated

Other Known Statements — (1974-1976)

11/14/1974: Report to NRC of 5.4 pCi/l (correction of 23.1 pCi/l reported on 6/1972 for Point Beach — correction reported as <10 in Errata Sheet, Annual Survey Report for 1975.)

3/1975 — Report to LAND for non-nuclear sites: monthly averages down to <5 pCi/l

4/1975 — Report to State Senator: <5 pCi/l

12/1975 — Report to LAND: "<10 pCi/l could be noted as '0' or not detectable."

TABLE VI

ILLUSTRATED VARIATIONS IN ANNUAL AVERAGE CONCENTRATIONS OF IODINE-131 (pCi/l) AS EFFECTS OF VARIOUS STATED DETECTION LIMITS. (DL) LA CROSSE MILK MONITORING: 1970-1971[a]

Date	Range of Weekly Reports: High	Low	Monthly and Annual Averages Computed by: Given Value[b]	<5 DL[c]	<10 DL
1970					
Jan	5.88	-2.6[d]	3.67	5	
*Feb	7.09	-0.37	3.91	<5	
*Mar	6.0	1.3	3.76	<5	
*Apr	9.92	0.79	3.89	<5	
May	37.75	10.32	19.26	20	
June	12.9	8.6	10.46	9	
*July	7.81	3.3	3.45	<5	
Aug	7.81	5.68	6.74	10	
*Sept	14.3	2.41	7.19	<5	
Oct	5.98	1.81	3.78	5	
Nov	10.44	-6.10	6.3	5	
*Dec	15.36	-1.07	4.81	<5	
Annual Averages:			6.46	4.8	2.5
1971					
Jan	5.73	3.7	4.47	5	
*Feb	5.28	-0.32	2.92	<5	
Mar	10.29	.25	5.43	5	
Apr	7.22	4.04	5.2	5	
May	16.67	3.47	8.47	8	
June	9.87	4.38	7.19	7	
July	7.81	3.05	4.80	5	
Aug	11.48	1.71	6.1	6	
*Sept	8.13	1.56	4.48	<5	
Oct	4.36	1.31	3.09	<5	
Nov	12.85	1.49	7.14	7	
*Dec	8.28	-12.65	3.9	<5	
Annual Averages:			5.26	4.0	0.0

[a] Concentrations refer to reports for La Crosse station in computer printouts for non-nuclear monitoring stations by Radiation Protection Section, Wisconsin Dept. of Health and Social Services. Dose estimates in this study are based on additional reports. See Study Design.

[b] Computed by LAND/LEAF Research Team from computer printouts received 10/1976.

ERRATA: Page 61, top - type missing:

[c] Computed by Radiation Protection Section for LAND/LEAF, 2/1975.

[d] Minus (-) concentration reports--indicating "less than background."

* Months for which a <5 Detection Limit average negates **ALL** presence of I-131 for the month. A <10 Detection Limit negates all presence of I-131 in all except two months for 1971-1972.

DL indicates "detection limit."

Interpretation of "detection limits" can have very significant effects on ultimate dose estimates. The Wisconsin laboratory has stated that concentrations below declared detection limits should be interpreted as 0 (zero) picocuries per liter. (8) ...

c Computed by Radiation Protection Section for LAND/LJ 3/1975.

d Minus (–) concentration reports – indicating "les ground."

* Months for which a <5 Detection Limit average i. presence of I-131 for the month. A <10 Detection Li. presence of I-131 in all except two months for 1971-1

DL indicates "detection limit."

Interpretation of "detection limits" can have very nificant effects on ultimate dose estimates. The Wiscon laboratory has stated that concentrations below declared detection limits should be interpreted as 0 (zero) picocu ies per liter. (8) Table VI illustrates a comparison of this interpretation to a straight acceptance of concentrations as recorded. In all cases, interpretation of "less than detectable" levels as 0 results in lower annual average concentrations than averaging of exact results (2 to 5.26 picocuries lower). In fact a "less than" 10 pCi/l "detection" level eliminated all trace of I-131 for one annual average. There can be no question about the actual presence of I-131 in milk during the 24 month period examined since at least one test result for 23 months was above 5 pCi/l and at least one test result for 9 months was above 10pCi/l (range from 0 to 37.75 pCi/l).

The U.S. Environmental Protection Agency (EPA) has concluded that when "the objective is to realistically estimate the dose" the "best estimate values . . . are preferable" to trying to estimate the dose from concentration records which assume a particular "detection" level. (9) In the R3 Diet Dose methodology all calculations are based on the best estimate exact picocurie levels of contamination. While the precision of measurement may be doubtful, it is obvious that a "realistic" dose estimate cannot be attempted by assuming that concentrations below a stated "detection level" must be interpreted as meaning that absolutely NO Iodine-131 is present.

While I-131 concentrations in milk up to 10 pCi/l are well below federal guideline levels requiring remedial action, they are significantly within the guidelines established by the NRC for limiting I-131 doses from nuclear reactors. (10) The NRC "design objective" goals would control the operation of nuclear reactors when the thyroid dose is above 15 millirems per year. An annual average of 5 picocuries per liter in milk will produce an internal diet thyroid dose of 15.84 millirems/year to a child and 28.2 millirems/year to an infant. (11) I-131 has a relatively short life (8 day half-life) and, therefore, except for short periods following nuclear testing, all of the I-131 detectable in milk will have originated from nuclear reactors and related facilities. Thus an I-131 detection limit of 5 to 10 pCi/l cannot demonstrate a reactor's compliance with "design objective" standards for continued operation. (12)

The results of an EPA cross-check evaluation of the laboratory cannot be applied to I-131 testing at these levels since none of the samples used had positive concentrations below 10 pCi/l. (13) However, the report by EPA following the Chinese nuclear testing in the fall of 1976 can be compared to the Wisconsin laboratory reported I-131 concentrations for the same period. (14) Figure III and Table VII present the comparison. Only one of the seven reports by the Wisconsin laboratory is in agreement with EPA results; six out of seven reports are below I-131 concentrations found by the EPA.

Cesium-137. The frequency of monitoring and fluctuation of a claimed "detection" level discussed under Iodine-131 apply as well to Cesium-137 testing. However, Cs-137 concentrations have ranged much higher than I-131 concentrations. (Annual averages -- Cs-137: .23 to 139.88 pCi/l; I-131: 2.755 to 6.99 pCi/l.) Since testing precision is more difficult at lower concentrations, the problem of accuracy of Cs-137 estimates is significantly applicable only to more recent years when Cs-137 levels were generally in lower ranges.

The difficulties of detection in recent years is illus-

TABLE VII

REPORTS OF I-131 IN MILK — OCTOBER 1-9, 1976
PICOCURIES PER LITER

Stations	Wisconsin Reports	U.S. EPA Reports
Rice Lake	1.9 (10/6/76)	20 – <50
Eau Claire	3.0 (10/8/76)	20 – <50
Green Bay	6.5 (10/6/76)	5 – <10
Wausau	2.6 (10/7/76) 4.1 (10/8/76)	10 – <20
Madison	2.5 (10/4/76) 4.4 (10/7/76)	5 – <10
La Crosse	2.0 (10/5/76)[a] 6.7 (10/5/76)[b] <1.2[c]	10 – <20

---

[a] Malin farm milk sample, Gamma-Scan

[b] Pedritti farm milk sample, Gamma-Scan

[c] Reported in Annual Report for La Crosse reactor, Radiochemical

References:

EPA, Assessment of Fallout in the United States from Atmospheric Nuclear Testing on September 26 and November 17, 1976 by the People's Republic of China. EPA-502/5-77-002, August 1977.

Radiation Protection Section, Department of Health and Social Services, Madison, Wisconsin, Annual Report: Environmental Radiation Survey — LaCrosse Boiling Water Reactor, January-December 1976.

Radiation Protection Section, Wisconsin, RADIOLOGICAL DATA SHEETS, as provided to and analyzed by Another Mother Fund for Peace Radiation Monitoring Research Staff.

Radiation Protection Section, Wisconsin, COMPUTER PRINT-OUTS OF MILK MONITORING.

FIGURE III

IODINE-131 MILK MONITORING BY U.S. EPA, OCTOBER 1-9, 1976

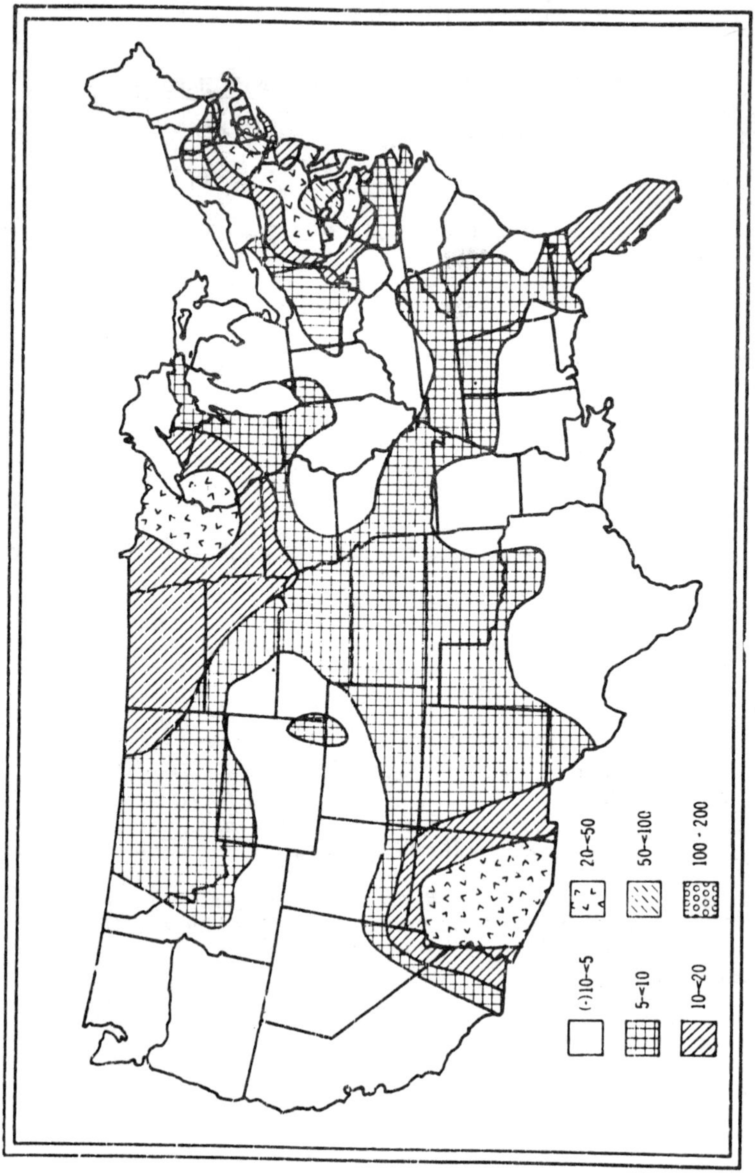

Distribution of iodine-131 in milk. Average concentrations October 1-9, 1976 (pCi/l).

Reference: U.S. Environmental Protection Agency, EPA Assessment of Fallout in the United States from Atmospheric Nuclear Testing on September 26 and November 17, 1976 by the People's Republic of China, August, 1977. (EPA-520/5-77-002), p. 33.

trated by the presence of negative values indicating "less than background" concentrations in the sample. Since this is impossible, the negative values reflect either analytical difficulties or setting an assumed "background" too high. These negative values do not appear in Cs-137 reports until 1972. In the period 1973-1976 most reported concentrations are negative values. (15)

The EPA cross-check on the Wisconsin Laboratory's ability to measure Cs-137 is applicable since the known values of samples in the cross-check do correspond to the range of values reported by the laboratory during the period 1963-1976. The cross-check reveals that the precision of the laboratory is in direct proportion to the amount of contamination in samples. For samples with over 100 pCi/l, underestimation was .1%; for 50 to 75 pCi/l, underestimation was 6.1%; for 24.9 pCi/l, underestimation was 16.5%; for 3 pCi/l, underestimation was 92%. Table VIII applies the cross-check analysis to the Wisconsin annual average concentrations of Cs-137 (1963-1976) as reported by the state laboratory. Hypothetical underestimation of Cs-137 concentrations in the range of 3 pCi/l to 24.9 pCi/l (not cross-checked by EPA) has been derived by applying the consistent tendency of reduced precision in direct proportion to concentrations in samples.

Table VIII shows that the reported Cs-137 levels are probably only slightly below actual contamination levels in the years 1963-1965; in 1966 the underestimation is probably about 16.5%; but for the years 1967-1976 reported levels may underestimate actual contamination by 50% to 92%. (16)

The dose estimates from Cs-137 in this study are based on reports as given by the state laboratory. Therefore it must be concluded that actual Cs-137 doses are much higher particularly for the period 1967-1976.

Strontium-90. Sr-90 concentrations in milk are generally reported monthly, however there are no reports for a significant number of months which may affect the accuracy of dose estimates. Table IX summarizes the most significant omissions — it does not include additional omissions primarily during winter months or isolated instances

TABLE VIII

PROBABLE UNDERESTIMATION OF CESIUM-137 DOSE IN WISCONSIN FROM EPA CROSS-CHECK OF WISCONSIN LABORATORY

Known Value EPA Cross-Check Sample (pCi/l)	Wisconsin Laboratory Value — Cross-Check Sample (pCi/l)	Percent Under-Estimation— Wisconsin Laboratory	Annual Averages Cs-137 in Wisconsin Milk—Wis. Lab Reports	
			pCi/l	Year
			.23	1976
			1.316	1973
			1.636	1975
3	.5	92%	3.063	1974
3	-2.8*		5.98	1972
(8.5)	**	(73.1%)	8.56	1968
			8.63	1969
			9.77	1967
			11.216	1970
			11.918	1971
(14)	**	(54.25%)		
(18.5)	**	(35.4%)		
24.9	20.3			
24.9	21.3	16.5%	24.97	1966
50	46.7			
50.3	45			
		6.1%	62.96	1965
75	71			
75	72.3			
80—206 (18 remaining samples)		.1%	125.95	1964
			139.88	1963

* Minus value interpreted as 0 (zero).
** No EPA Cross-Check samples in these ranges.
Numbers in parentheses ( ) indicate probability of underestimation at concentration levels not checked by EPA — as indicated by extension of trend to consistently improve detection capability as pCi/l levels increase.

References:

Computer Printout, Wisconsin State Laboratory, Milk Data, provided by George B. Morgan, Director, U.S. EPA Environmental Monitoring and Support Laboratory, Las Vegas, Nevada, to J.D. Littlejohn, Research Assistant, Another Mother Fund for Peace, January 11, 1978, as reported in Another Mother Fund for Peace, RADIATION MONITORING STUDY.

since these are not expected to affect dose estimates significantly.

The Table also shows months in which Sr-90 levels were exactly the same for several months in a row at one station or at two stations. Since such exact duplications of results are rare it appears that only one test was actually performed during these successive months.

The year most affected by these omissions is 1967 (only one test at one station reported). For purposes of dose estimate a steady decline in Sr-90 from 1966 to 1968 was assumed to derive average pCi/l for 1967. The years most affected by lack of testing during summer grazing months, when Sr-90 levels are expected to peak, are 1972, and 1963. Since all test results for 1972 for the Rice Lake and Eau Claire stations are exactly alike it is assumed that only one station was tested. (17)

The EPA cross-check on the analytical ability of the Wisconsin laboratory to measure Sr-90 is not useful since approximately 90% of the samples checked contained concentrations much higher than those usually encountered in Wisconsin milk. However, the EPA conclusion that "less than half" of all laboratories checked "are doing a satisfactory job of strontium-90 analysis" indicates a general lack of accuracy of test results. (18)

In this study the Sr-90 reports as given were used to derive dose estimates. The realism of the dose estimates is qualified by the number of actual test results reported. For all stations the chance of departures from reality is highest in the years 1972, 1967, 1966, 1963 and 1971.

The average annual concentrations of Sr-90, Cs-137 and I-131 for the six milk monitoring stations are presented in Tables X, XI, and XII.

TABLE IX

REPORTING OF Sr-90 IN MILK, pCi/LITER

YEAR	MADISON	GREEN BAY	WAUSAU
1963	------- No Report (NR) — All Stations -----------		
	NR-Aug	NR-Oct	*38-Oct *28-Nov
1964		29-Jan thru Sept	
	14-Oct thru Dec	*21-Oct thru Nov	*24-Oct thru Dec
1965	------- No Report (NR) — All Stations -----------		
	NR-Sept		
1966	NR-Aug	NR-Aug	NR-Sept
	------- No Report (NR) — All Stations -----------		
1967		NR-Jan	NR-Jan
	------- No Report (NR) — All Stations -----------		
1969			NR-Dec
1970			
1971	NR-Jan		
	5.8-Sept thru Dec	1.6-Sept thru Dec	2.2-Sept thru Dec
1972			
		NR-Mar	
	NR-Apr, May	NR-Apr, May	NR-Apr, May
	------- No Report (NR) — All Stations -----------		
			NR-Sept
1973	NR-Feb		
	NR-Aug-Sept		
1974	NR-Mar	NR-Mar	NR-Mar
1975			

* — Same reported pCi/l at two different stations for two or more successive months.

** — Range in pCi/l for Sept thru Dec was 1.4 to 5.6 in Annual Environmental Radiation Survey Report for LaCrosse reactor. Previously reported concentrations for LaCrosse "Reference Station" dropped from Annual Reports June 1971 — only "Genoa" reports continued.

*** — Computer printout reports for La Crosse non-nuclear testing station dropped June 1972.

	LA CROSSE (a)	LA CROSSE (b)	EAU CLAIRE	RICE LAKE
	— January through July, 1963 --------------------------			
	*38-Oct			
	*28-Nov			
	22-Jan thru June			
	*21-July thru Dec		*24-Oct thru Dec	41-Oct thru D
	— January through August, 1965 -----------------------			
	NR-Aug	NR-Aug	NR-Aug	NR-Aug
	— October through December, 1966 --------------------			
	NR-Jan		NR-Jan	NR-Jan
	— February through December, 1967 -------------------			
		NR-Aug-Sept	NR-Dec	
		NR-Feb		
	NR-July	NR-Aug		
		NR-Nov	NR-Oct	
	.70-Sept thru Dec	**	1.8-Sept thru Dec	3.6-Sept thru D
			*10.1-Jan	*10.1-Jan
			* 3.0-Feb	* 3.0-Feb
			* 3.0-Mar	* 3.0-Mar
	NR-Apr, May		NR-Apr,May	NR-Apr, May
	--June through August, 1972 --------------------------			
	***		* .10-Sept	* .10-Sept
			* 4.7-Oct	* 4.7-Oct
			* 3.6-Nov	* 3.6-Nov
		NR-Dec	* 6.3-Dec	* 6.3-Dec
		NR-Apr		
		NR-July		
		NR-Dec		
1975				NR-Mar
		NR-June		
		NR-Aug		
		NR-Oct		

NR — Indicates No Report

(a) — Computer Printouts for non-nuclear LaCrosse testing station
(b) — Annual Environmental Radiation Survey Reports — LaCrosse reactor.

TABLE X

STRONTIUM-90 — Picocuries per Liter in Milk

Year	Rice Lake	Eau Claire	Wausau
1963	35.2	34	32.2
1964	37.66	28.03	25.83
1965	45.02	24.12	31.32
1966	29.8	25.9	29.63
1967*	21	17.9	20.9
1968	12.2	9.91	12.2
1969	9.58	11.6	8.65
1970	9.06	10.6	6.72
1971	8.71	6.06	7.56
1972	4.40	4.40	3.86
1973	9.83	11.0	6.58
1974	10.3	16.6	10.9
1975	11.0	10.7	6.66
1976	9.53	11.5	8.42

* No Sr-90 monitoring reports for Rice Lake, Eau Claire, Wausau, Green Bay testing stations. Figures for 1967 derived from average of 1966 and 1968 concentrations based upon assumptions of continuing decline in Sr-90 during this year.

Figures for Madison and LaCrosse stations are for one report from each station for 1967.

LaCrosse	Green Bay	Madison	Average
30	20.4	16	28
21.5	27.6	18.1	26.45
31.44	17.9	25	29.1
22.26	16.3	21.7	24.3
11.3	11.8	9	15.3
7.58	7.42	5.5	9.1
7.71	5.82	4.7	8.0
7.04	5.83	5.4	7.44
6.74	4.30	5.7	6.5
4.33	2.45	3.0	3.74
5.46	6.38	6.7	7.66
5.22	11.32	7.1	10.24
7.5	6.32	5.4	7.9
7.33	6.04	4.9	7.95

Annual averages computed from computer printouts of concentrations prepared by Radiation Protection Section, Wisconsin Department of Health and Social Services for all non-nuclear testing stations.

Computer printout for LaCrosse station terminated 7/1/72. Monitoring by Radiation Protection Section for LaCrosse reactor, May 1965 through December 1976, in addition to available computer printouts used to derive LaCrosse averages.

TABLE XI

CESIUM-137 – Picocuries per Liter in Milk

Year	Rice Lake	Eau Claire	Wausau
1963	174.33	155.9	155.57
1964	149.05	127.5	148.0
1965	75.45	62.7	79.65
1966	27	24.6	34.63
1967	11.5	10.2	13.52
1968	11.95	8.6	15.1
1969	10.7	9.9	15.08
1970	13.35	11.8	15.47
1971	15.48	13.3	14.44
1972	8.74	5.85	6.07
1973	1.99	1.06	1.79
1974	4.44	2.95	3.34
1975	2.71	1.09	2.03
1976	.18	.32	.47

Annual averages computed from monthly averages derived from computer printouts of concentrations prepared by Radiation Protection Section, Wisconsin Department of Health and Social Services for all non-nuclear testing stations.

LaCrosse	Green Bay	Madison	Average
128.19	124.12	101.19	139.88
113.62	115.86	101.67	125.95
48.07	59.58	52.32	62.96
19.81	21.75	22.05	24.97
7.72	9.02	6.66	9.77
4.87	7.71	3.13	8.56
5.44	6.91	3.77	8.63
8.49	10.66	7.53	11.216
10.44	10.85	7.0	11.918
6.65	4.74	3.85	5.98
1.19	1.53	.34	1.316
2.53	3.05	2.07	3.063
1.9	1.51	.58	1.636
0	.26	.15	.23

Computer printout for LaCrosse station terminated 7/1/1972. Monitoring by Radiation Protection Section for LaCrosse reactor, May 1965 through December 1976, in addition to available computer printouts used to derive LaCrosse averages.

TABLE XII

IODINE-131 — Picocuries per Liter in Milk

Year	Rice Lake	Eau Claire	Wausau
1963	2.43	2.07	4.01
1964	3.42	3.36	2.86
1965	6.75	6.18	5.4
1966	6.13	7.05	8.67
1967	5.82	3.83	5.42
1968	6.12	6.16	6.98
1969	3.94	4.45	4.88
1970	7.65	6.32	7.35
1971	5.36	6.46	5.35
1972	3.87	3.27	4.15
1973	2.98	4.05	2.79
1974	3.97	2.28	2.08
1975	3.04	3.39	2.43
1976	2.52	3.61	2.71

Annual averages computed from monthly averages derived from computer printouts of concentrations prepared by Radiation Protection Section, Wisconsin Department of Health and Social Services for all non-nuclear testing stations.

LaCrosse	Green Bay	Madison	Average
2.19	3.27	2.56	2.755
2.74	3.37	3.68	3.238
5.67	4.47	5.55	5.67
7.25	5.45	7.41	6.99
4.9	5.68	3.78	4.905
5.8	5.18	5.66	5.98
5.0	5.57	5.75	4.93
5.7	6.37	7.18	6.76
6.4	4.78	5.10	5.075
3.7	3.77	4.13	3.815
3.5	3.39	3.67	3.398
2.4	3.27	3.99	2.998
2.2	3.08	3.06	2.866
3.3	3.68	3.13	3.158

Computer printout for LaCrosse station terminated 7/1/1972. Monitoring by Radiation Protection Section for LaCrosse reactor, May 1965 through December 1976, in addition to available computer printouts used to derive LaCrosse averages.

Local and Regional Exposure Patterns

The radiation dose to individuals in the Wisconsin case study will be influenced by omission from the R^3 methodology of four major radiation pathways due to the lack of precise data: Doses from radionuclides other than Strontium-90, Cesium-137 and Iodine 131; doses from inhaled nuclides; doses from drinking water; and, doses from dairy products not included in the R3 Diet calculations. (See discussion of Adjustments to Total Internal Dose, in Section A: Methodology.)

Of these major omissions, the consumption of additional dairy products is of particular significance since Wisconsin is a principal milk-producing state. The relative availability, variety and inexpensiveness of dairy products will encourage higher consumption rates. Dairy desserts, not included in the R3 Diet, are popular and readily available in Wisconsin, particularly to children and young adults. Cities, as well as rural areas, feature a great variety of choices. The promotion of dairy products in the state encourages higher consumption rates. Varieties are abundant in desserts, yogurts, cheeses. A particularly popular form of cheese, fresh cheddar curds ("nuggets") is consumed fresh and will, therefore, contain concentrated short-lived radionuclides (such as Iodine-131) in addition to the long-lived nuclides in aged cheddar cheese. Fresh cheddar curds are not included in the R3 Diet. On farms, milk, butter and cream are used in more cooking since they are relatively inexpensive and abundant (i.e., in cream pies, puddings, soups). Since farm families tend to consume fresh milk, Iodine-131 and other short-lived nuclides will be more significantly consumed than indicated by concentrations in "store-life"(4days) sample analyses.

The dairy component of the fission dose is calculated for cow's milk but, increasingly, goats are prized for superior nutrition and the convenience of smaller yield for people not in the dairy business. Goat milk will yield five times as much Iodine-131, 17.5 times as much Strontium-90, and 25 times as much Cesium from their food as cow's milk. (19) It is estimated that individuals in Wisconsin who substituted goat milk for cow milk (1963-1976) will have accumulated a fission diet dose four times higher than the case-study population. (Calculation for fluid milk consumption only; goat cheese consumption not included.)

Radiation dose from Drinking Water (a major omission from dose calculations) can only be surmised since statewide monitoring data for the period is not available. A New York City survey shows a ratio of Sr-90 in water to Sr-90 in milk of one to 20. (20) The U.S. EPA assumes a drinking water consumption rate of two liters/day; compared to milk consumption of one liter/day for growing children in the Wisconsin case study, one could expect the drinking water Sr-90 dose to equal one-tenth of the milk Sr-90 dose if the New York City ratio prevailed. It is estimated that 6% of the total diet Sr-90 dose in New York City in 1966 was due to drinking water. (21) On the other hand, Drinking Water was judged to be "the major source of discharged radionuclides to populations of Richland, Kennewick and Pasco, Washington" near the Hanford, Washington reactors. (22)

In Wisconsin drinking water from deep wells will probably contain less fission products than water from reservoirs, streams, lakes and shallow wells. The Mississippi River and Lake Michigan are receiving bodies for liquid reactor wastes. Lakes and extensive marshlands can become reservoirs of fission products from land runoff and airborne desposits of radionuclides. Runoff from irrigated lands has been recognized by the state Department of Natural Resources as a significant source of fertilizer and pesticide pollution of streams, lakes and shallow wells. (Designated as "non-point" pollution.)

Swimming areas in the state are often closed in the summer because of high bacteria counts due to barnyard and septic tank pollution. The potentially large source of barnyard runoff as a contribution of radiation in drinking water and other foods is not included in any pathways models in current use by regulatory agencies. In current Nuclear Regulatory Commission Environmental Impact statements for proposed nuclear plant_, for instance, the figures depicting pathways of radiation to

FIGURE IV

FLOW CHART: STRONTIUM IN THE COW

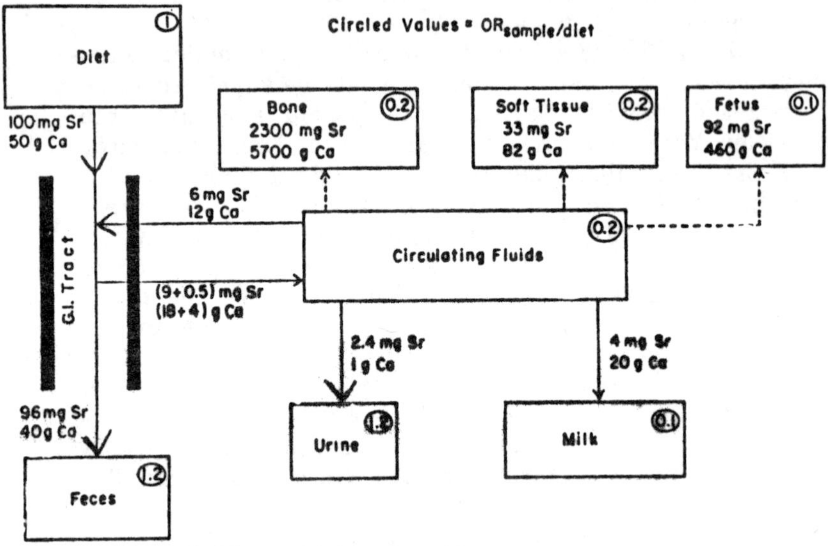

Typical flow and balance for stable strontium (mg) and calcium (g) in the cow.
Values beside arrows represent daily amounts and those within boxes the total amounts in tissues.

Reference: R. Scott Russell, ed., Radioactivity and the Human Diet, (Oxford, New York: Pergamon Press, 1966), p. 271.
Elongation of arrows indicating radiation pathways of animal wastes containing strontium to man added.

FIGURE V

RADIATION PATHWAYS

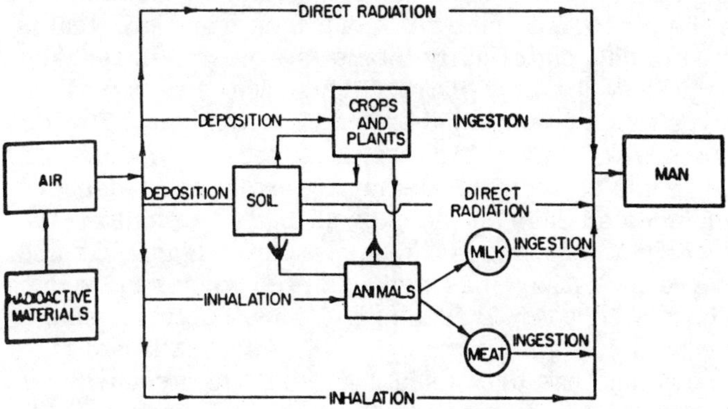

Pathways between radioactive materials released to the atmosphere and man (Environmental Protection Agency, 1972).

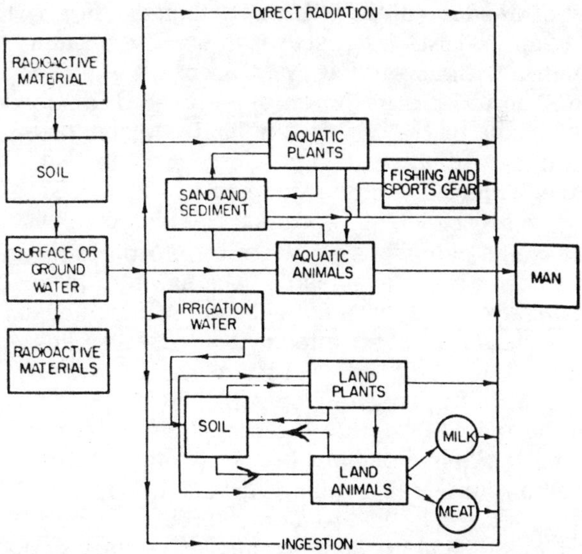

Fig. 17-2. Pathways between radioactive materials released to ground and **surface** waters (including oceans) and man (**Environmental Protection Agency, 1972**).

Reference: Merril Eisenbud, Environmental Radioactivity, 2nd Ed., (New York: Academic Press, 1973), pp. 235-236.

Elongation of arrows indicating radiation pathways of animal wastes containing strontium and other radionuclides added.

humans from the pictured cow include only milk. (23) However, R. Scott Russell, reviewing the flow of Strontium-90 ingested by a cow through food, shows that milk will contain only one-tenth of the ingested Strontium-90 while the feces and urine will contain one and two-tenths. Ninety-six percent of daily intake is excreted in feces. See Figure IV for flow chart. Figure V shows radiation pathways figures published by the Environmental Protection Agency in 1972. Note that these earlier models, no longer in use do indicate clearly the cycle of re-deposition of ingested radionuclides by animals back into soil from where the pathways to humans through animals and crops are repeated. It is well established by research that "only a very small fraction of many nuclides is absorbed from the gut [of cows]." (24) As a matter of fact, this knowledge has been used in the past to discourage the public from believing that Strontium-90 dose could be reduced by cutting milk consumption. Merril Eisenbud points out that supplying human calcium needs through cow's milk will produce about one-tenth the Sr-90 dose as a supply of needed calcium through plants. "In short, the lactating animal has been shown to be a strontium decontaminator with an efficiency of about 90%." (25) In 1966 the Atomic Energy Commission used this "decontaminating" virtue of the cow to emphasize that: "the study of milk as an indicator food does not mean that milk is a preponderant factor in determining total Sr-90 intake. As a matter of fact, because a dairy cow puts 10 times as much calcium as strontium into the milk in proportion to the amounts ingested, the calcium in milk will be the least contaminated of all food sources of calcium . . . . It is usually futile to try to reduce radioactivity intake significantly by alterations in diet. An excellent example is presented by milk." (26)

It must be pointed out that the corollary to this "decontaminating" efficiency of the cow, in providing relatively strontium-free calcium to humans, is the concentrating efficiency in producing strontium-rich barnyard effluent. The grazing animal efficiently gathers radioactive fallout from a wide area, concentrates it in feces and urine, and leaves it where it is much more accessible to humans than it would have been in the fields. Ninety percent of a long-lived radionuclide thus concentrated cannot be simply dismissed from the pathways of radiation to humans.

The direct pathways of this concentrated radiation to humans are: barnyard runoff into streams, lakes and groundwater, eventually entering drinking water and aquatic foods; concentrated exposure through use as fertilizer in consumed grains, vegetables and fruits (27); direct contamination of barnyard workers; and, re-contamination of milk. In fact, Garner points out that insoluble rare earths, Ru and Pu-239 found in cow's milk can only be accounted for through animal feces contamination of milk in the milking shed. (28)

The presence of chemicals and minerals in drinking water consumed by some individuals can enhance the radiation hazards. For instance, recent research has revealed that chlorination of drinking water may enhance the biological hazards of plutonium in the water. (29) Drinking water in Wisconsin could also be contaminated by leaching from the environmental sink created by extensive marshlands where long-lived radionuclides from past and present fallout will concentrate.

These considerations lead to the conclusion that Wisconsin Drinking Water, excluded for lack of monitoring data from these dose calculations, could contribute an additional fission dose of from 6% to 20% of the total diet intake.

In addition to being subject to higher than normal fission fallout from bomb testing by virtue of latitude and rainfall patterns, Wisconsin's topography is conducive to storing of the contamination in environmental reservoirs where it remains a potential recontamination hazard to humans. The state has extensive marshlands and moss or lichen ground cover which are well recognized as efficient radiation pathways. Russell points out that the highest concentrations of radiation are found in mosses, lichens, marshes and shallow water. (30) Most radionuclides are found in much higher concentrations in marshy plants than in dryland plants. Concentrations ranged from 1.8 to 8.6 times higher in marsh-grown sphagnum moss for Zn-65, Mn-54, Ce-144, Zr and Nb-95 and Cs-137 than for dryland willow leaves. (31)

Cesium-137 (a gamma emitter) is of special concern in marsh, lichen and moss deposits where it remains for very long periods and can be transmitted to humans through edible plants, meat from animals who eat the plants (berries or nuts), as well as inhalation and direct external gamma exposure. Research reviewed by Garner concludes that 95% of Cs-137 deposits on lichens are retained — the "weathering half-life" may be as long as thirty years — concentrations from 1961 to 1965 increased by a factor of three to four. Appreciably higher concentrations in moss compared to lichens were found for Mn-54, Sr-90, Ru-106, Cs-137 and Ce-144. (32)

Wisconsin's heavy snowfalls may enhance cesium deposits since they appear to be promoted by snow melt. An examination of samples from the vicinity of the Dresden (Illinois) nuclear power plants by the Bureau of Radiological Health revealed Cs-137 concentrations in snow of 10 to 15 picocuries per liter while in 7 rainwater samples only one, registering 0.4 picocuries per liter, had concentrations above the level of detection. Strontium-89 concentration was more than 10 times as high in snow samples near the meteorological tower than at a distance; Sr-90, Zr-95, Ru-106, Ba-140, and Ce-141 concentrations were also higher in snow samples near the tower. (33)

The most significant pathway to Wisconsin residents from foods affected by the marsh—moss topography is probably venison. It is estimated that 150,000 deer were harvested by hunters in 1978. Poaching is a serious problem to game wardens and the actual harvest could be much higher. (34) Venison is a staple in the diets of many families — freezers are stocked in the fall for a season's supply. Central Wisconsin, where the deer harvest is most abundant, is also the milk-monitoring station with above average Cs-137 concentrations — highest concentrations in 7 out of 14 years. Northwestern Wisconsin (with highest milk Cs-137 levels also in 7 out of 14 years) is the least populated and most rural area where residents are more likely to consume venison in quantity.

Venison has long been recognized as an especially

high contributor to radiation dose. In addition to the generalizations already discussed in Section A, the following facts are worth noting: 25% of the Cs-137 deposits on lichen and moss are absorbed by deer. (35) Lapps in Northern Finland consuming reindeer were found to have radiation doses much higher than Lapps in Southern Finland eating other meats: liver dose, 11.3 times higher; bone dose, 2.5 times higher; gonad dose, 12.1 times higher. In the New York state Nuclear Fuel Reprocessing area the EPA concluded that the whole body dose commitment to a "maximum" individual consuming 50 kilograms of venison per year would be 151 times higher than to an individual consuming only 2 kilograms of venison per year. (36)

The Wisconsin State health department has published no radiological data on Wisconsin venison, but the U.S. Department of Health, Education and Welfare identified concentrations of 69 pCi/kg Cs-137 in deer muscle in Illinois, 40 km north of the Dresden plant which is close to the southern border of Wisconsin. (Cs-137 concentration of 90 pCi/kg were identified in deer only 7 km from the Dresden plant.) (37) The Cs-137 concentration in milk in southern Wisconsin (Madison station) in the same year (1969) was only 3.77 pCi/liter – 18.3 times lower than in deer muscle. The Madison station Cs-137 milk was the lowest in the state – 4 times lower than Central Wisconsin and 3 times lower than Northwest Wisconsin. If this ratio of cesium-deer/cesium-milk is applicable to the rest of the state, it would follow that Central Wisconsin venison would have had about 275 pCi/kg Cs-137 in 1969.

Concentrations of Cs-137 in white tail deer muscle in 1968 ranged from 100 to 100,000 pCi/kg. (38) But Cs-137 is not the only radionuclide of concern in venison. Strontium-90 as well as many other radionuclides not included in the Wisconsin dose estimates contribute substantially to the dose. In the West Valley, NY study, one-third of the dose from venison was attributed to other radionuclides. Cs-134 is one of the most important – it is produced in nuclear reactors in a ration of 2: Cs-134/3: Cs-137, but it is not monitored by the State. (39)

Garner notes that cesium concentrations increase

greatly as it moves up the food chain. 'Nine-fold increases have been reported through the plant-mule deer-cougar chain." (40) Thus, the deer acts as an efficient concentrator of cesium in the diet of humans. It is difficult to estimate the influence of this factor on the subjects of the Wisconsin case study in the absence of monitoring. However, there is no doubt that individuals consuming deer meat, particularly in Central and Northern Wisconsin, will have doses substantially higher than derived from the R3 Diet which excludes this pathway. A significant number of individuals, considering an annual deer harvest of at least 200,000, will be affected in varying degrees.

Individuals who consume relatively more pork and poultry will also be exposed to higher Cs-137 doses. The percentages of daily intake concentrated in muscle of pork, poultry and beef respectively are 26%, 450% and 4%. (41) It is thought that swine and, to an even greater extent, poultry pick up appreciable amounts of Cs-137 from soil. Thus chickens given free access to soil and grass have cesium flesh levels of 1.1 to 1.3 times higher than those given access only to commercial feed. (42)

Iodine-131, Strontium-90 and Cesium-137 are also found in eggs. Fresh eggs, of course, will have a higher concentration of short-lived I-131, which is transferred primarily to egg yolks, than eggs purchased commercially. The transfer of Iodine-131 from feed to eggs is 8 times higher than from feed to cow's milk. Consumption of fresh eggs daily can therefore substantially increase the I-131 dose particularly to children's thyroids. Transfer of Cs-137 is 1.6 to 2.5 times higher to eggs than to cow's milk. (43) Fourteen radionuclides were found to be transferable to eggs with I-131 and Cs-134 (not monitored in Wisconsin) in highest concentrations. (44)

The relatively higher radionuclide concentrations in pork, poultry and eggs will produce higher fission doses to Wisconsin farm families, suburbanites or urban families who prefer "farm-fresh" products. The popularity of Wisconsin-produced sausages will add particularly to whole-body doses from cesium.

Wisconsin-produced fruits and vegetables will also

yield higher fission doses than calculated depending upon relatively higher proclivity to concentrate nuclides, areas in which they are produced and relative proportion to total diet (often determined by availability and expense).

Table XIII summarizes the most important agricultural products of Wisconsin (by rank among states). Table XIV shows the Sr-90 and Cs-137 concent of some of these agricultural products which are likely to be consumed in larger-than-normal quantities due to their general availability in Wisconsin. Publication of radionuclide content of fruits and vegetables is sparce and no data has been published by the state of Wisconsin. However, the federal agency data in Table XIV will provide an estimate of the possible content of major fruits and vegetables relative to radionuclide content of Wisconsin milk during the same period. Almost all fruit and vegetable concentrations noted are higher than milk. Exact location of fruit and vegetable samples is not given in the literature for most analyses and very few are identified as Wisconsin products; it is noted that "the highest levels of strontium-90 were observed in the Central States." (45) The data summarized in Table XIV is for the highest findings. It is not intended, here, to represent average concentrations for dose estimate purposes but, rather, as a guide to the relative potential radioactive contamination of various foods compared to monitored milk contamination.

The high levels of radioactivity in some of the foods on Table XIV is due to the fact that they grow in the marshy – moss habitat. (See discussion of venison.) Very high contamination levels are noted for cranberries, a major crop in Wisconsin. Cranberries are cultivated in wetlands and are irrigated and protected from frost by flooding. It has been noted that "flooding may render available to plants nuclides which would be virtually unavailable from soil." (46) Data for many of these nuclides, such as Co-60, Ce-144, Mn-54, is not available for cranberries. Freshwater plants concentrate radioactivity more than sea water plants due to relatively low mineral levels of fresh water. The range of increased concentration of cesium in freshwater plants over that present in water is 80 to 4,000; for Mn, 1300 to 600,000. (47)

Strawberries and raspberries rank next to cranberries for radiation concentration of cesium and strontium. Both are widely cultivated in Wisconsin; access is plenti-

TABLE XIII

HOW WISCONSIN RANKS AMONG THE STATES
IN AGRICULTURE, 1973

Commodity	% of U.S. Production	Rank Among States	1973 Production
DAIRYING			
Butter	19.5	2	179,406,000 lbs.
Cheese, all (excluding cottage)	39.9	1	1,071,041,000 lbs.
American	43.1	1	720,104,000 lbs.
Blue	59.5	1	17,711,000 lbs.
Brick	73.8	1	17,060,000 lbs.
Italian	35.3	1	199,674,000 lbs.
Limburger	66.8	1	1,728,000 lbs.
Muenster	78.2	1	35,720,000 lbs.
Swiss	20.0	2	32,764,000 lbs.
Condensed milk, bulk			
Sweetened, skim	67.8	1	48,614,000 lbs.
Sweetened, whole	78.0	1	60,086,000 lbs.
Unsweetened, skim	15.7	2	127,012,000 lbs.
Unsweetened, whole	26.3	1	41,987,000 lbs.
Dry products			
Buttermilk	20.0	2	8,573,000 lbs.
Malted milk powder	100.0	1	16,263,000 lbs.
Skim milk for animals	4.6	4	446,000 lbs.
Skim milk for humans	9.7	3	89,242,000 lbs.
Whey	46.4	1	357,365,000 lbs.
Whole milk	5.8	6	4,527,000 lbs.
Evaporated whole milk, unsweetened, case [1]	....	...	....
Ice Cream	2.1	16	16,006,000 gal.
Milk production	16.0	1	18,442,000,000 lbs.
CROPS			
Apples	0.8	21	50,000,000 lbs.
Barley	0.2	27	777,000 bu.
Beans, lima for processing	7.0	3	6,900 tons
Beans, snap for processing	18.6	2	137,600 tons
Beets for canning	30.2	2	60,600 tons

Commodity	% of U.S. Production	Rank Among States	1973 Production
Cabbage, kraut	25.4	2	57,550 tons
Cabbage, total	11.6	2	1,621,000 cwt.
Cherries, tart	2.7	6	2,400 tons
Corn for grain	3.1	8	173,470,000 bu.
Corn for silage	9.7	1	10,710,000 tons
Corn, sweet for processing	23.5	2	510,500 tons
Cranberries	36.4	2	760,000 bbl.
Cucumbers for pickles	8.4	4	51,050 tons
Hay, alfalfa	11.0	1	8,613,000 tons
Hay, all other	3.6	8	2,009,000 tons
Hay, all	7.9	1	10,622,000 tons
Maple Syrup	9.8	3	84,000 gal.
Oats	8.5	5	56,170,000 bu.
Onions, commercial	1.3	[illegible]	392,000 cwt.
Peas, green for processing	24.8	[illegible]	121,950 tons
Peppermint for oil	3.5	[illegible]	112,000 lbs.
Potatoes	3.9	9	11,515,000 cwt.
Rye	0.7	25	176,000 bu.
Seed, red clover	7.8	6	2,250,000 lbs.
Seed, timothy	3.0	7	700,000 lbs.
Soybeans for beans	0.4	24	6,025,000 bu.
Strawberries	0.8	12	38,000 cwt.
Tobacco	1.1	10	18,963,000 lbs.
Wheat, all	0.1	37	890,000 bu.
OTHER PRODUCTS			
Eggs	1.9	20	1,267,000,000 eggs
Wool	0.6	26	822,000 lbs.
Honey	4.8	5	11,440,000 lbs.
LIVESTOCK			No. on Farms January 1, 1974
Cattle and calves, all	3.4	9	4,400,000 head
Milk cows and heifers that have calved	15.9	1	1,796,000 head
Chickens [2]	1.7	20	6,821,000 birds
Hogs and pigs, all [2]	2.6	12	1,575,000 head
Turkeys raised [3]	3.2	10	4,229,000 birds

---

[1] Data not shown when less than 3 plants reported or individual plant operations might be disclosed.

[2] December 1, 1973 [3] 1973

Source: Wisconsin Department of Agriculture, Statistical Reporting Service, October 1974.

Reference: The State of Wisconsin, 1975 Blue Book, p. 568.

TABLE XIV

STRONTIUM-90 AND CESIUM-137 CONTENT OF FOODS, 1962-63
Highest Levels Reported, U.S.

Food	Picocuries per kilogram Sr-90	Cs-137
Cranberries	110	562
Blueberries	83	372
Strawberries	59	
Cherries		236
Root vegetable	494	
Turnip	42	413
Potatoes	157	135
Squash	34	
Grains	1340	
Wheat	495	561
Onions	70	
Mustard Greens		3990
Broccoli	244	
Cabbage	9	1030
Radishes	92	
Beets	104	
Navy Beans		1934
Carrots	64	
Soybeans	34	674
Soybean Flour	25	336
Nuts	543	
Oats	447	
Sweet Potatoes	494	
Cheese	416	
***Dairy Product	1040	
**Cheese(Switzerland) (1973)	101.7	141
*Wisconsin milk (1963-76)	61.3	305.3
Animal Feed	6860	

References:

U.S. Department of Health, Education and Welfare, Radiological Health Data Reports, December 1962; February 1963; November 1964 — except for the following:

*Computer Printouts of Milk Monitoring, Radiation Protection Section, Wisconsin Department of Health and Social Services;

**U.S. Environmental Protection Agency, Office of Radiation Programs, Radiation Data and Reports, Vol. 15, No. 10, October 1974.

***Not specifically identified as "cheese," but highest concentration of milk would be expected in cheese.

ful at "pick-your-own" fields and fresh fruit stands. Blueberries are commonly gathered from their wild habitat.

Leafy green vegetables will have the highest concentrations of short-lived Iodine-131 — increased dose will be in proportion to relative freshness. They are also the highest among vegetables in Sr-90 content.

Potatoes were notably high in Sr-90. Potato cultivation is a major industry in Wisconsin and it would not be unreasonable to assume a consumption level approaching half of fluid milk for a sizeable sample of the population. Some ethnic groups consume considerable quantities in a variety of forms; poorer populations find potatoes locally purchased to be economic; young people consume extra snack potatoes (chips and french fries).

Potatoes are cultivated in central Wisconsin on large acreages of sandy soils and spray-irrigated as are crops of other vegetables. Under these circumstances Cs-137 becomes a particular problem. Even in the absence of fresh fallout of Cs-137 it can easily be resuspended by wind or mechanical disturbance or redeposited on plants by splashing rain or irrigation water. In fact, resuspension from accumulated reservoirs "may become the predominant source of this radionuclide to animals and man." (48) Unlike other radionuclides, Cs-137 is freely translocated throughout the plant and appears in root vegetables as well as leafy ones. Resuspension was found to be "the major route" of plant contamination in the Montebelle Islands near Australia — also near the Nevada test site; and 30% of Cs-137 fallout on hay in Colorado was from resuspended soils. Central Wisconsin's sandy soils, extensively cultivated and unprotected from winds, may partially account for the fact that this milk monitoring station has the highest Cs-137 dose despite the fact that it ranks only third among stations for Sr-90 dose. Sr-90 tends to be absorbed into soils more readily and is not as available for resuspension.

Extensive sand and gravel mining operations and building projects may also account for resuspension of cesium deposits. As a gamma emitter, Cs-137 can also produce external radiation doses to animals and humans.

Radionuclides other than Cs-137 are also a resuspension hazard. Deposits of Pu-239 can become disintegrated to respirable size by weathering and, resuspended, can produce lung doses. (None of the alpha-emitting radionuclides are included in the R3 dose estimates.) Grazing cattle have been dosed by Pu-239 in this manner. (49)

Spray irrigation has been found to result in high concentration of H-3 (tritium) in crops. Tritium is released from Pressurized Water reactors at the rate of 5 Ci/year – MW(e). (50) Tritium in Wisconsin foods is not analyzed in any State publications.

The concentrations of Zn-65 in pasture grasses spray irrigated with Columbia River water was 400 times higher than in the water and 25 times higher than in milk and beef. (51) The Hanford (Washington) reactors release effluents. No reports for Zn-65 in Wisconsin foods are available in State publications.

Soybeans have the highest reported concentration of Cs-137 excepting navy beans and greens. They are widely cultivated in Wisconsin for both human and meat and milk animal consumption. The increase in nutrition consciousness in Wisconsin, as well as nationwide, leads to increased consumption of soybeans and soybean flour as well as other "health" foods which can give above average radiation doses to a particularly vulnerable segment of the population: the young or those already in poor health attempting to improve their health through various "health" foods.

Bone meal, for instance, can deliver a sizeable Sr-90 dose since animals, as well as humans, concentrate most of this nuclide in bone and bone marrow where it is absorbed with calcium. In fact, the "high levels of Strontium-90 that could be contained in bone particles" was recently cited by U.S. District Judge William B. Bryant as an argument against a proposal by the Department of Agriculture to allow bits of bone and bone marrow in ground meats. (52)

Whole grain cereals and breads are also higher in Sr-90 than in finer milled products which removes part of the outer Sr-90 deposits. Doses could be much higher for those who raise their own wheat and consume it with minimum milling. Bread in the USSR using flour of lower extraction had a Sr-90 content almost twice the finer milled bread in New York City. (53) Buckwheat has almost twice the Sr-90 found in wheat. The especially high Sr-90 concent of whole wheat bread has been known a long time. (54)

Seeds of all kinds (sunflower, flax, etc.), valued by the nutrition-conscious, are high concentrators of Sr-90. In fact, it has been noted that "oils and fats derived from oilseeds and oil bearing fruits . . . could be an important source of long-lived nuclides to both domestic animals (as cattle cake) and man (margarine)." (55)

Freshness of produce of all kinds is valued by the health-conscious and others. Higher short-lived radionuclide doses will be unavoidable. It has also been noted that cooking of vegetables and meats removes some nuclide contamination but that consumption of vegetable and meat liquids by particular ethnic groups or health food enthusiasts will negate any benefit of cooking. (56)

In an agricultural area such as Wisconsin, where farm-fresh produce is available to urban as well as rural populations, the time factor allowance for decay of short-lived radionuclides assumed in dose estimate calculations is probably overestimated. The transport time (from time of harvest to diet) assumed by the Nuclear Regulatory Commission, for instance, is: 14 days for fruits, grains, vegetables; 4 days for milk; 20 for meat and poultry; 7 for sports fish. Food consumed by farm families and many others who normally grow their own vegetables or are sportspeople or purchase at farm markets across the state will have doses from short-lived nuclides which are not monitored or accounted for in dose estimates.

Variety in soil conditions and cultivation practices can influence levels of radionuclides in fruits, vegetables, grains and meat-and-milk producing animals. For instance, poorly fertilized pastures will transfer 50 percent more of deposited Iodine-131 and Strontium-90 to grazing animals than well fertilized and limed pastures. However, calcium-poor soils would require application of "several tons of lime per acre" to lower Strontium-90 uptake. (57) Livestock grazing in marshy or mossy areas will be subject to the same problem of high uptake of Cesium-137 as deer. If "wild hay" cut for silage is obtained from marshy areas, it, too, will have higher Cs-137 levels than hay grown on well fertilized fields, or corn silage. Plants with "rough, coarsely grooved, hairy leaves retain more" radionuclides; thus, highest retention of Sr-90 and Cs-137 occur consecutively in dense grass, potatoes, wheat, cabbage and sugar beets. (58) No State data is available on radionuclide content of animal feed but the U.S. Health, Education and Welfare Department found by far the highest concentration of Sr-90 in the early 1960's in livestock feed (6,860 pCi/kg) and the highest levels in all foods occurred in the central states. (59) Long-lived radionuclides can remain for many years on surface layers of permanent pastures or in the plowed layers of cultivated soils. (60)

Some individuals or groups in the population may have accumulated higher radiation doses due to external or inhalation pathways which are not detailed in these dose calculations. Such doses are most obviously due to occupational exposure. Monitored nuclear power plant workers in Wisconsin had an average accumulated whole body dose of up to 6,000 millirems by 1975. (61) The annual average dose to monitored workers in 1975 was: LaCrosse, 1,420 mrems; Point Beach, 1,300 mrems; Kewaunee, 500 mrems. The additional dose to these workers from a Wisconsin R3 Diet has not been considered by regulatory agencies in limits imposed on worker radiation exposure.

Physicians, radiologists, radiation researchers will have occupational exposure to fission products. Other occupations may also produce higher doses. It has been suggested that "insoluble material trapped in the inflourescences of cereal plants could pose an inhalation hazard through air contamination during operations such as threshing and milling." (62) Farmers and farm families could also absorb additional external exposure from contact with barnyard feces where radiation can be concentrated.

Summary: Factors Affecting Realism of Wisconsin Fission Dose Estimates

The fission dose findings presented in the following pages of this study are a best-estimate for the average individual for that portion of the dose which clearly can be quantified on the basis of radionuclide monitoring in Wisconsin milk by the State and expanded to include other internal and external doses in a ratio established as average for the Northern Latitude or for the United States.

The major adjustments to these best-estimate dose findings for Wisconsin are the following:

For all residents: Dose estimates will be underestimated by omission of radiation exposure from Drinking Water; Dairy desserts, butter and cream; Inhalation; and, doses from unmonitored radionuclides. In addition, the apparent underestimation of Cesium-137 in monitored milk could result in significant underestimation of whole-body dose. (63)

For particular individuals and groups: Significantly higher-than-average doses will accrue to: Nuclear power plant workers; farm families and others consuming fresh produce or exposed to barnyard runoff; individuals whose diet is high proportionately in goat milk, venison, "health" foods, potatoes, pork, poultry, berries or fresh green vegetables. As Garner points out: "dietary habits" can have a "profound effect . . . on the total intake of a radionuclide and the consequent radiation exposure." (64)

The additional exposure to individuals in an important consideration in estimating dose. In the first place, it is <u>individuals</u>, not <u>populations</u>, who will actually suffer the health effects attributable to radiation exposure — to the <u>individual</u> who has developed bone cancer due to radiation, the fact that the <u>population</u> exposure is not expected to result in high increased risk of bone cancer is a meaningless calculation. Furthermore, the <u>genetic</u> consequences of excessively high exposure will affect not only the exposed individual but his or her population group as well for generations into the future. As Dr. John Gofman observes: if "members of the public received <u>no dose at all</u>, there would still be a serious introduction of 'person-millirems' from the occupational exposure of workers in the nuclear power industry itself, that is, considering only the workers of reproductive age. The genetic effect upon humans from such an introduction of 'person-millirems' will be just the same as if everyone were exposed to a smaller dose leading to the same number of 'person-millirems.'" (65)

In interpreting the Wisconsin dose findings, then, it must be kept in mind that these quantifications are, to some degree, underestimated for all individuals and significantly underestimated for some individuals and groups.

Findings: Wisconsin Realistic Fission Dose

All fission dose estimates in this case study are based upon milk monitoring data as recorded by the Radiation Protection Section, Wisconsin Department of Health and Social Services. They do not include expected additional doses to all subjects in varying degree as discussed in the foregoing section. The Dose Estimates would, of course, be more realistic if these dose adjustments were included. However, it is difficult to quantify the doses from even those sources which will affect all individuals since the State has not collected monitoring data. The dose methodology developed here quantifies only that part of the dose which can be derived directly from actual monitoring of selected radionuclides. This element of conservatism should be kept in mind in assessing the dose quantifications.

First, the 50-year Dose Commitment for the R3 Diet was calculated. The R3 Diet Dose includes only that portion of a total fission dose which accrues from ingestion of foods containing Strontium-90, Cesium 137 and Iodine-131. It does not include doses from other radionuclides, external radiation or inhalation. The R3 Diet Dose cannot, of course, represent the whole dose expected. It does represent that part of the dose which is, without question, evident from monitoring data.

The R3 Diet Dose Commitments to various organs and total body for the period 1963-1976 in the six milkshed regions are presented in Table XV. Doses are indicated for two groups: (1) A Growing Child — accumulation to a child born in 1963 to age 15; (2) An Adult — accumulation since 1963 to the present for persons over 18 years of age in 1963. Table XVI summarizes the State average R3 Diet Dose Commitment to all organs for the two population groups.

Doses to a Growing Child are a Realistic Lifetime Dose Commitment beginning at birth. (The pre-birth dose to the fetus has not been calculated.) Adult doses are only for that part accumulated from 1963-1976. Since these Adults would be expected to have absorbed an additional dose commitment during the pre-1963 atomic era, the Adult doses represent only a part of a Realistic Lifetime Fission Dose.

The 14-year accumulated dose commitments to all organs and total body are higher for the growing child population than for the adult: Growing Child Total Body – 1055.77 millirems; Adult -- 491.98 millirems. Growing Child Bone – 4249.13 millirems; Adult – 1896.68. A factor of 3.34 difference for Thyroid dose: Growing Child –189.18; Adult -- 55.28.

Table XV shows a wide range of dose, also, among the six State monitoring area. Bone, Total Body, and GI-LLI doses are almost twice as high in the Northwest region (Rice Lake, Bone Dose – 5665.51 millirems) as they are in the Southeastern region (Madison, Bone Dose – 3158.1). (Adult doses vary in the same degree as the Growing Child doses above.) In general, the 14-year accumulation is higher at Rice Lake, Eau Claire and Wausau (all above state average) than at LaCrosse, Green Bay and Madison (all below state average). Dose accumulations to Liver, Lung and Kidney, which are due to Cs-137 only, show less divergence between the state areas than Bone, Total Body and GI-LLI doses, which are due primarily to Sr-90 ingestion. (The Wausau station is highest in accumulated dose commitments from Cs-137). Thyroid dose accumulations show very little divergence among stations indicating more equalized Iodine-131 pollution of the whole state. (The lack of divergence for thyroid doses may also reflect the difficulty of precise measurement at the reported levels of concentration.)

The greatest degree of divergence in R3 Accumulated Dose Commitments is seen between dose to an Adult in

TABLE XV

ACCUMULATED R^3 DIET DOSE COMMITMENTS: WISCONSIN, 1963-1976
TO GROWING CHILD AND ADULT (mrems)*

Testing Station	Bone	Liver	Total Body	Thyroid	Lung	GI-LLI	Kidney
Rice Lake:							
Child	5665.51	199.3	1412.053	191	22.8	79.66	62.7
Adult	2501.185	68.019	646.12	57.185	7.64	72.2	23.3
Eau Claire:							
Child	4769	170.78	1185.547	178.3	19.3	69.1	53.2
Adult	2194.39	58.44	566.28	51.466	6.565	63.37	20.04
Wausau:							
Child	4722	202.1	1170.905	206.6	23.1	66.1	63.8
Adult	2095.87	67.7	546.52	57.185	7.61	60.5	23.215
LaCrosse:							
Child	3854.1	141.004	958.461	183.26	16.079	54.209	44.43
Adult	1732.84	48.16	448.13	51.466	5.41	50.04	16.55
Green Bay:							
Child	3326.5	149.6	823.266	183.7	17.1	47.5	47.2
Adult	1487.6	50.67	389.15	57.185	5.69	43	17.43
Madison:							
Child	3158.1	124.5	784.403	192.2	14.1	44.2	39.5
Adult	1368.25	41.9	355.69	57.185	4.7	39.5	14.46

*See Note for TABLE XVI.

STATEWIDE AVERAGE: ACCUMULATED R^3 DIET DOSE COMMITMENTS
WISCONSIN, (1963-1976) GROWING CHILD AND ADULT (mrems)*

	Bone	Liver	Total Body	Thyroid	Lung	GI-LLI	Kidney
Growing Child:							
Accumulated Dose	4249.13	164.55	1055.77	189.18	18.75	60.13	51.8
Average Committed per year	303.5	11.75	75.41	13.51	1.34	4.3	3.7
Adult:							
Accumulated Dose	1896.68	55.81	491.98	55.28	6.27	54.77	19.17
Average Committed per year	135.48	3.99	35.14	3.95	0.45	3.91	1.37

* A Realistic Lifetime Fission Dose Commitment was calculated for a child born in 1963: Infant dose — 1963; Child dose 1964-1973; Teen dose 1974-1976. Total accumulated lifetime commitment to a 15-year-old. Base: Milk consumption of 365 liters per year.

Adult Dose Commitment represents accumulation of 14 years (1963-1976) which has been added to the dose commitment prior to 1963. Base: Milk consumption of 310 liters per year.

"R3 Diet Dose Commitment" is ONLY for INGESTION of Sr-90, Cs-137 and I-131. It EXCLUDES doses from all other radionuclides and from Inhalation and External sources and all consideration of Dose Adjustments discussed herein.

the South-Southeastern Station (Madison) and dose to a Growing Child in the Northwestern Station (Rice Lake): difference by a factor of 4.14 for Bone dose and 3.97 for Total Body dose.

Persons born at the beginning of the atomic age in Wisconsin (in the early or mid-'50's), who will have received Growing Child doses during the height of the weapons fallout (1957-1964), will have accumulated dose commitments higher than those indicated for either of the population groups examined. Monitoring data prior to 1963 is not available for assessing dose to this population group in particular. A best-estimate is that this group will have an accumulated dose commitment equal to 70 to 80 percent of the sum of the Adult and Growing Child doses represented here.

Tables XVII and XVIII list 50-year Dose Commitments to Total Body from the R3 Diet only for Growing Child and Adult respectively, for each year. The yearly Growing Child doses reflect the changing diet intake and biological impact of radionuclides from Infant to Child to Teen. Thus the lower 1963 dose commitment reflects the lower diet intake of the Infant than for Child in 1964-1973. The 1974-1976 doses are for a Teenager — the drop in yearly dose commitment in 1974 does not indicate a drop in radiation levels but the relatively smaller biological impact of radionuclides on the body of a teenager compared to a Child. (Age-specific dose differences were derived from Nuclear Regulatory Commission Regulatory Guide 1.109.)

The Adult dose commitments by year (Table XVIII) do reflect real increases and decreases in monitored radiation levels since factors of diet intake and biological impact do not change for this population group.

Generally dose commitments are seen to decrease from a high in 1963-1965 (yearly commitments of 74.71 to 78.675 millirems to Adults) to a low in 1970-1972 (low of 9.41 millirems state average Adult dose commitment in 1972). In the 1973 to 1976 period dose commitments increase in varying degrees among stations to levels comparable to pre-1970 years. The state average yearly dose commitment in 1974 (24.59 millirems) is higher than the 1968 dose (22.45 millirems) for Adults.

The estimated Total Fission Dose Commitments, which consist of the basic R3 Diet dose plus other internal and external dose commitments derived by ratio from United Nations Scientific Committee on the Effects of Atomic Radiation reports for the Northern Latitude, are summarized in Table XIX for Total Body dose and Table XX for the Bone for the six monitored stations. (See methodology development for Total Fission Dose Commitment, Section A.) Since the Total Fission dose estimates are not based on actual state monitoring records for nuclides other than Sr-90, Cs-137 and I-131 or for External doses, they are not absolutely verifiable. However, in the absence of State monitoring, they represent the best-estimate calculation based on expected ratio of known to unknown radiation dose in the State, as indicated by the United Nations reports. Total Fission Dose Commitments to Total Body are slightly more than double the dose from R3 Diet only; slightly under double for Bone.

Again, the Total Fission dose commitments reflect wide divergence between Growing Child and Adult doses (Total Body: 1147.2 millirems to 2391.93 millirems). The increase by a factor of more than 2 over doses from the R3 Diet only is due primarily to external dose from gamma emitters (such as Cs-137) from fallout deposits. In the absence of sizeable fresh fallout it has been predicted that external dose from previous Cs-137 deposits will constitute a major part of total body dose. Other doses, not included in the R3 Diet, which contribute relatively less to the Total Fission dose are internal doses from H^3, C^{14} and Fe^{55} and external doses from a variety of Short-lived fission products.

Accumulated Total Fission Dose commitments to Bone are by far the highest of any dose calculations in this study: (See Table XX) Range: Adult (Madison — 2701.2 millirems; to, Growing Child (Rice Lake) — 11,085.8 millirems. The primary source of Total Fission excess over R3 Diet only for Bone dose are the Short-lived radionuclides. The ratio of external Short-lived dose to internal Sr-90 dose concluded by United Nations reports for the Northern Latitude cannot be either verified or discounted for the state of Wisconsin since the state does not publish a record of these. In the presence of fresh Sr-90,

TABLE XVII

TOTAL BODY R^3 DIET DOSE COMMITMENTS:
WISCONSIN 1963-1976 (mrems) TO GROWING CHILD*

Year	Rice Lake	Eau Claire	Wausau
1963	80.97	78.0	74.0
1964	253.998	190.154	177.265
1965	296.779	160.49	208.279
1966	194.915	169.481	194.322
1967	136.853	116.678	136.341
1968	79.857	64.793	80.067
1969	62.795	75.832	57.061
1970	59.601	69.478	44.580
1971	57.476	40.158	49.955
1972	28.096	28.903	25.419
1973	63.834	71.354	42.76
1974	32.08	51.376	33.847
1975	34.108	33.062	20.68
1976	29.387	35.47	25.988
Accumulations:			
Sr^{90}	1386.089	1163.31	1144.078
Cs^{137}	25.66	21.92	26.486
I^{131}** Residual	.304	.317	.341
Totals	1412.053	1185.547	1170.905
Average per yr.	100.86	84.68	83.64

* See Note for TABLE XV.

LaCrosse	Green Bay	Madison	Average
68.73	47.14	37.2	64.38
146.9	186.59	124.08	179.83
206.9	119.974	165.5	192.99
145.5	107.08	142.094	158.9
73.72	77.069	58.765	99.9
49.33	48.598	35.85	59.75
50.36	38.176	30.709	52.49
46.17	38.492	35.496	48.97
44.4	28.589	37.404	43.0
28.544	16.193	19.698	24.48
35.48	41.466	43.441	49.72
16.29	35.12	22.039	31.79
23.242	19.593	16.687	24.56
22.59	18.636	15.114	24.53
940.09	803.499	767.96	1034.17
18.066	19.457	16.117	21.28
.311	.31	.326	.318
958.467	323.266	784.403	1055.77
68.46	58.80	56.03	75.41

** Iodine-131 Residual is not included in yearly Dose Commitment figures.

TABLE XVIII

TOTAL BODY R[3] DIET DOSE COMMITMENTS:
WISCONSIN 1963-1976 (mrems) TO ADULT*

Year	Rice Lake	Eau Claire	Wausau
1963	98.87	94.41	90.11
1964	102.50	77.74	74.31
1965	113.53	62.78	81.37
1966	73.15	63.66	73.40
1967	50.89	43.40	50.82
1968	30.03	24.29	30.30
1969	23.69	28.42	21.86
1970	22.69	26.21	17.32
1971	22.05	15.56	19.22
1972	11.22	10.96	9.70
1973	23.52	26.22	15.79
1974	24.85	39.69	26.18
1975	26.37	25.51	16.00
1976	22.65	27.34	20.04
Accumulations:			
Sr^{90}	601.59	528.02	502.18
Cs^{137}	44.43	38.17	44.24
I^{131}** Residual	.10	.09	.10
Totals	646.12	566.28	546.52
Average per yr.	46.15	40.45	39.04

---

* See Note for TABLE XV.
** See Note for TABLE XVII.

LaCrosse	Green Bay	Madison	Average
82.48	59.32	46.86	78.675
61.01	75.70	51.90	73.86
78.88	47.73	63.96	74.71
54.61	40.62	53.47	59.82
27.52	28.82	21.96	37.235
18.43	18.30	13.33	22.45
18.79	14.43	11.49	19.78
17.46	14.78	13.49	18.66
16.92	11.16	14.15	16.51
10.86	6.24	7.47	9.41
13.07	15.28	15.94	18.30
12.62	27.16	17.04	24.59
17.98	15.14	12.88	18.98
17.41	14.37	11.65	18.91
416.60	355.97	328.25	455.44
31.44	33.08	27.34	36.45
.09	.10	.10	.096
448.13	389.15	355.69	491.99
32.01	27.80	25.41	35.14

TABLE XIX

ACCUMULATED TOTAL FISSION DOSE COMMITMENT: TOTAL BODY WISCONSIN, 1963-1976, GROWING CHILD AND ADULT (mrems)*

	Rice Lake	Eau Claire	Wausau	LaCrosse	Green Bay	Madison	Average
Growing Child:							
Accumulation	3195.84	2683.58	2656.31	2169.92	1868.59	1777.39	2391.93
Average Committed per year	228.27	191.68	189.74	154.00	133.47	126.96	170.85
Adult:							
Accumulation	1502.34	1315.75	1278.85	1042.71	912.51	831.06	1147.20
Average Committed per year	107.31	93.98	91.35	74.50	65.18	59.36	81.94

* Total Fission Dose Commitment is the basic R^3 Diet Dose plus other Internal and External Doses as derived by perdicted Ratio from UNSCEAR reports of fission dose in the Northern Latitude. See Table II. Total Fission Dose Commitment does not include Dose Adjustments discussed herein.

TABLE XX

ACCUMULATED TOTAL FISSION DOSE COMMITMENT: BONE WISCONSIN, 1963-1976, GROWING CHILD AND ADULT (mrems)*

	Rice Lake	Eau Claire	Wausau	LaCrosse	Green Bay	Madison	Average
Growing Child:							
Accumulation	11085.8	9510.0	9468.1	7689.4	6679.6	6315.2	8458.0
Average Committed per year	791.84	679.3	672.3	549.24	477.11	451.1	604.0
Adult:							
Accumulation	4928.7	4322.7	4141.8	3415.7	2942.6	2701.2	3742.0
Average Committed per year	352.0	308.76	295.8	243.98	210.2	192.9	267.0

* Total Fission Dose Commitment is the basic R^3 Diet Dose plus other Internal and External Doses as derived by predicted Ratio from UNSCEAR reports of fission dose in the Northern Latitude. See Table II. Total Fission Dose Commitment does not include Dose Adjustments discussed herein.

as R3 Dose commitments document, in Wisconsin in recent years one would expect concommitant external dose from fresh Short-lived radionuclides. Other internal doses to bone have been omitted from all dose calculations — Strontium-89, for instance, is produced by nuclear reactors in larger quantities than Strontium-90. Sr-89 has a half-life of only 50.5 days compared to 30 years for Sr-90. Eisenbud notes that it was shown in 1962 "that for any dose of Sr-90 incorporated into bone during the first year of a child's life the Sr-89 may deliver as much as 3.4 times the dose delivered by Sr-90." (66) Sr-89 is not included in any Wisconsin dose estimates.

The Total Fission Dose Commitment accumulations to Total Body and Bone do not include any of the additional sources of fission radiation dose expected to have been accumulated by all Wisconsin individuals nor additional doses to groups or individuals as discussed previously.

The Fission Dose Commitments, discussed above and represented in Tables XV through XX, will accrue over a 50-year period before the continued fissioning and decay of radionuclides within body cells is completed. For some radionuclides this biological effective half-life is relatively short: Cesium-137 will have delivered all of the Dose Commitment within 16.4 months after the initiation of radiation impact on the body; I-131, a short-lived nuclide, is absorbed quickly by the body. However, Strontium-90, stored in bone and bone marrow, continues its persistent radiological impact on the body over a 50-year decay period. The 50-year Dose Commitment from Sr-90, therefore, is much larger than the Annual biological effect on body cells in any one year. The biological effects will continue to be delivered, year after year within the body, for many years following the initial year of ingestion.

Table XXI summarizes the accumulation by December 1977 of this Annual Dose impact from radionuclide dose initiated in the 1963-1976 period. Figures are for Total Body dose from the R3 Diet only for both Growing Child and Adult population. A wide range of divergence is again noted from accumulated annual dose of 143.54 millirems to an Adult (Madison) to 535 millirems to a

Growing Child (Rice Lake). The average annual dose in any one year ranges from 14 millirems/year to 38.21 millirems/year.

Table XXII summarizes the 14-year accumulation of Annual doses to Bone for all stations and the two population groups. The average yearly dose delivered to bone ranges from 35.9 millirems/year to an Adult (Madison) to 157.6 millirems/year to a Growing Child (Rice Lake). The dose to Bone is from the R3 Diet only and is primarily due to ingestion of Strontium-90. It does not include external dose or possible dose from Strontium-89 or other internally ingested radiation. It can, therefore, not represent the whole accumulation of Annual doses to Bone, from radiation initiated in the 1963-1976 period.

Tables XXI and XXII discussed above are for the 14-year accumulation of Annual doses from the R3 diet only. The Annual Total Body dose for each year (1963-1976) for each monitoring station and the State average accumulation as it accrues year by year is presented in Table XXIII for the Growing Child and in Table XXIV for the Adult population. The highest Child doses have been delivered in the very recent years as a dose continues to be delivered every year from all past ingestion and a new dose is initiated each year. The range of Annual Total Body dose to a Growing Child is from 1.57 millirems (Infant Dose, 1963, Madison) to 45.39 millirems (Teen Dose, 1974, Rice Lake). Except for 1963 (a low initial dose average of 6.36 millirems including no previous dose commitments) the Adult Annual Total Body doses do not vary as much from year to year as for the Growing Child. This is a reflection of the much greater impact of Cesium-137 dose biologically on the Adult than on the Child and less impact from Strontium-90 dose. From 1964 to 1976 the Adult Annual dose varies from 9.51 millirems to 22.07 millirems.

Most of the Dose Commitment is left after annual doses through 1976 — 58.8% for Adults and 62.6% for Growing Child. This radiation dose from 1963-1976 ingestion of three radionuclides will continue to be delivered every year within the body through year 2026 — resulting in a biological impact every year until radionuclides have completed their decay process and been eliminated from

TABLE XXI

TOTAL BODY ACCUMULATED ANNUAL DOSE ABSORBED
FROM R3 DIET DOSE COMMITMENT, 1963-1976 — WISCONSIN (mrems)*

	Rice Lake	Eau Claire	Wausau	La Crosse	Green Bay	Madison	Average
Growing Child.							
Accumulation	535.002	425.539	444.549	362.535	309.492	294.202	395.22
Average Annual Dose	38.210	30.400	31.750	25.900	22.110	21.010	28.23
Adult:							
Accumulation	261.54	216.82	225.13	182.79	157.86	143.54	197.95
Average Annual Dose	18.68	15.49	16.08	13.06	11.28	10.25	14.00

* Portion of 50-year Dose Commitment which will have completed biological impact by December 1977.

TABLE XXII

ACCUMULATED ANNUAL DOSES ABSORBED: BONE
FROM R3 DIET DOSE COMMITMENT, 1963-1976 — WISCONSIN (mrems)*

	Rice Lake	Eau Claire	Wausau	La Crosse	Green Bay	Madison	Average
Growing Child:							
Accumulation	2206.3	1764.1	1852.5	1500.4	1296.1	1221.4	1640.0
Average Annual Dose	157.6	126.0	132.3	107.2	92.6	87.2	117.2
Adult:							
Accumulation	934.7	769.6	785.6	651.6	545.1	503.3	698.0
Average Annual Dose	66.8	55.0	56.1	46.5	38.9	35.9	50.0

* Portion of 50-year Dose Commitment which will have completed biological impact by December 1977.

TABLE XXIII

ANNUAL TOTAL BODY R3 DIET DOSE
DELIVERED 1963-1976 (mrems) WISCONSIN GROWING CHILD

Year	Rice Lake	Eau Claire	Wausau
1963	3.07	2.93	2.80
1964	16.44	14.19	14.10
1965	33.44	25.09	26.32
1966	40.06	27.47	31.23
1967	41.60	31.23	34.92
1968	43.09	32.39	36.45
1969	43.41	33.17	36.99
1970	44.10	35.44	37.36
1971	45.09	36.54	37.76
1972	45.21	35.92	37.85
1973	44.47	36.09	36.97
1974	45.39	38.23	37.47
1975	44.50	38.20	37.06
1976	44.17	37.83	36.34
Accumulations: ----			
I 131 **Residual	.304	.317	.341
Total Delivered	534.344	425.037	443.961
50-year Commitment	1412.053	1185.547	1170.905
Percent Commitment left to be delivered	62.16%	64.15%	62.08%

* State of Wisconsin — Average Accumulations

** Iodine-131 Residual is not included in yearly Dose Commitment figures.

LaCrosse	Green Bay	Madison	Accumulated Average
2.50	1.99	1.57	2.48
11.97	11.44	8.77	15.30
22.08	22.24	18.10	39.85
27.20	21.96	22.52	68.26
28.40	23.27	24.15	98.86
28.08	24.03	22.93	130.02
28.60	24.42	22.95	161.61
29.79	24.96	23.53	194.14
30.72	25.69	24.49	227.52
31.13	25.37	24.76	260.89
30.70	25.02	24.66	293.88
30.64	26.25	25.55	327.80
29.95	26.36	24.95	361.30
29.96	25.77	24.59	394.41
.311	.310	.326	.30
362.031	309.080	293.846	394.70
958.461	823.266	784.403	1055.77
62.22%	62.46%	62.54%	62.60%

Note: R^3 Diet Dose is ONLY for INGESTION of Sr-90, Cs-137, and I-131. It EXCLUDES doses from ALL other radionuclides and from Inhalation and External sources and all consideration of Dose Adjustments discussed herein.

TABLE XXIV

ANNUAL TOTAL BODY R^3 DIET DOSE
DELIVERED 1963-1976 (mrems) WISCONSIN ADULTS

Year	Rice Lake	Eau Claire	Wausau
1963	7.95	7.24	7.13
1964	21.78	19.38	19.68
1965	22.07	17.79	19.34
1966	20.43	15.18	17.37
1967	18.66	14.66	16.37
1968	18.57	14.45	16.13
1969	18.58	14.61	16.29
1970	18.79	15.48	16.35
1971	19.25	15.94	16.46
1972	19.10	15.46	16.13
1973	18.33	15.07	15.31
1974	18.69	16.12	15.66
1975	19.24	17.28	16.23
1976	19.46	17.39	16.11
Accumulations:			
I^{131} ** Residual	.10	.09	.10
Total Delivered	261.00	216.14	224.66
50-year Commitment	646.12	566.28	546.52
Percent Commitment left to be delivered	59.60%	61.83%	58.90%

* State of Wisconsin — Average Accumulations

** Iodine-131 Residual is not included in yearly Dose Commitment figures.

LaCrosse	Green Bay	Madison	Accumulated* Average
6.06	5.42	4.38	6.36
16.47	15.17	12.30	23.82
15.53	15.37	12.85	40.98
14.18	12.14	11.79	56.16
13.14	10.80	10.80	70.23
12.46	10.53	9.67	83.87
12.45	10.55	9.51	97.54
12.94	10.75	9.82	111.56
13.39	11.15	10.26	125.97
13.42	10.82	10.24	140.17
12.97	10.34	9.93	153.83
12.84	11.00	10.34	167.94
13.06	11.78	10.65	182.65
13.39	11.64	10.64	197.42
.09	.10	.10	.096
182.39	157.56	143.28	197.60
448.13	389.15	355.69	491.99
59.30%	59.51%	59.72%	58.80%

Note: R^3 Diet Dose is ONLY for INGESTION of Sr-90, Cs-137 and I-131. It EXCLUDES doses from ALL other radionuclides and from Inhalation and External sources and all consideration of Dose Adjustments discussed herein.

the body. The percentages of R3 Dose Commitment to be delivered in the future are not exactly the same for all monitoring stations since those stations with the highest doses in more recent years relative to other stations will have had little time to deliver an Annual dose to this date. The state areas in order of the highest to lowest percentage of Adult Annual Whole Body dose left for future delivery are: Eau Claire, Madison, Green Bay, La Crosse, Rice Lake and Wausau. 2.07 percentage points separate the extremes. The significance of this difference will be discussed later.

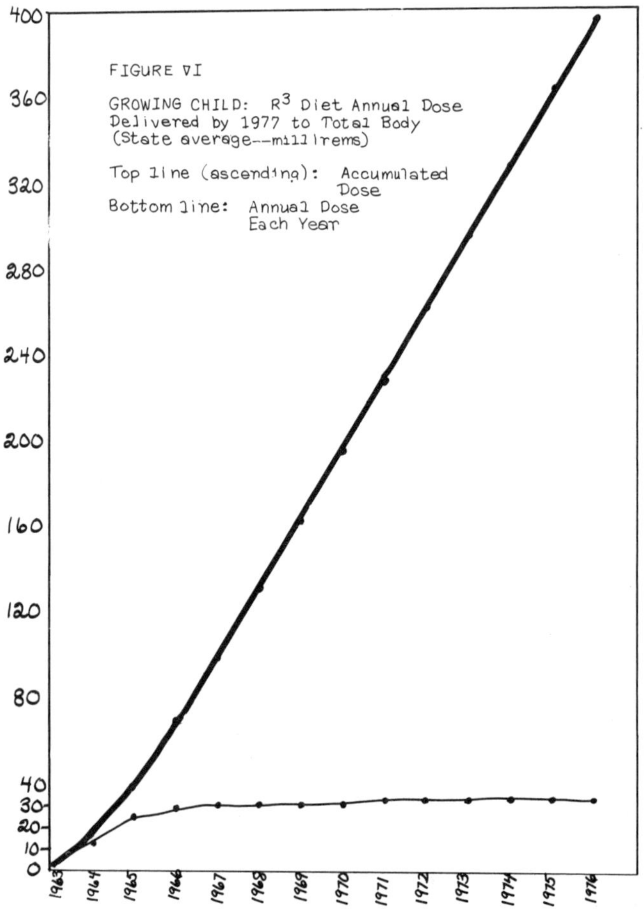

Figure VI and VII present a visual interpretation of the process by which the Annual dose delivered each year contributes to the mounting Annual Dose Accumulation. Note the much steeper slope of ascending dose accumulation to Growing Child than Adult. The dose accumulation line will continue to climb with every year even if no more new radiation is initiated in coming years. With appreciable new radiation the slope will become steeper. In any year, the end of the sloping accumulation line marks the biological impact by which expected current health effects are determined.

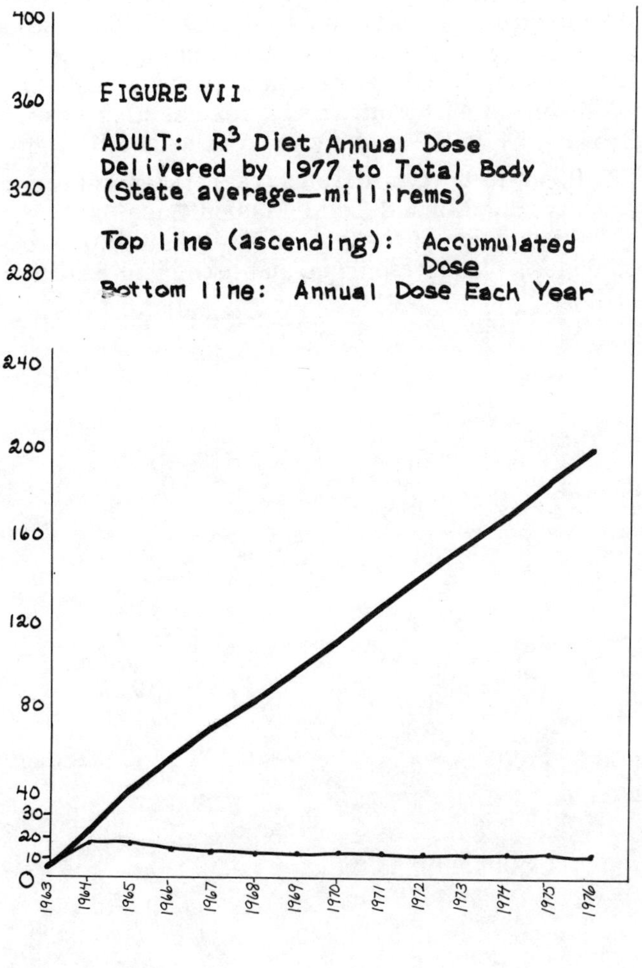

The year by year Annual doses to Total Body were estimated only for the R3 Diet dose (from ingested Sr-90, Cs-137 and I-131). Annual Total Fission Doses (including other internal plus external doses) would be expected to be approximately twice the R3 Diet dose as an average. The magnitude of Total Fission Dose over the R3 Diet dose for any one year is difficult to quantify without year to year monitoring records of other radionuclides or external dose. However, an estimated Accumulated Annual Total Fission Dose (1963-1976) was calculated with ratios for estimated external radiation derived from U.S. Environmental Protection Agency estimates for the 1963-1976 period in the United States. (See Table IV.)

The estimates for Total Fission Dose delivered by 1977 to Growing Child and Adult (total body dose) are summarized in Table XXV. The average Annual Total Fission dose to Adults ranged from 23.2 millirems/year (Madison) to 42.35 millirems/year at Rice Lake. For the Growing Child the range was from 47.6 millirems/year (Madison) to 86.6 millirems/year (Rice Lake) for total body average Annual Total Fission Dose.

Discussion of these dose findings relative to expected doses, trends, sources and estimated health risks follows.

TABLE XXV

TOTAL BODY – ACCUMULATED ANNUAL DOSES ABSORBED
FROM TOTAL FISSION DOSE COMMITMENT, 1963-1976 – Wisconsin (mrems)*

	Rice Lake	Eau Claire	Wausau	LaCrosse	Green Bay	Madison	Average
Growing Child:							
Accumulation	1212.9	964.7	1007.8	821.9	701.6	666.9	896.0
Average Annual Dose	86.6	68.9	72.0	58.7	50.1	47.6	64.0
Adult:							
Accumulation	592.95	491.6	510.4	414.4	357.9	325.4	448.8
Average Annual Dose	42.35	35.1	36.5	29.6	25.6	23.2	32.0

* Portion of 50-year Dose Commitment which will have completed biological impact by December 1977.

Discussion of Dose Findings

Realistic Fission Dose (R^3 Methodology-derived) for Wisconsin Compared to Expected (Model-derived) Fission Dose Estimates. In the Introduction to this study, it was postulated that model-derived dose estimates for large populations could not serve as accurate guides to the actual fission dose accrued by real individuals in a particular population.

Table XXVI compares the 1963-1976 accumulated dose commitment from ingestion of Sr-90 and Cs-137 in Wisconsin to the United Nations Scientific Committee on Effects of Atomic Radiation (UNSCEAR) estimate of dose commitments to the Northern Latitude population from ingestion of Sr-90 and Cs-137 from all bomb-testing prior to 1971. (It is conservatively presumed in this study that Wisconsin dose commitments for 1971-1976, excluded from the UNSCEAR report, will not be larger than Wisconsin dose commitments prior to 1963, excluded from Wisconsin calculations. The EPA has not noted any major rise in radiation levels in the United States from bomb testing since 1971.)

The Total Body dose commitment to a Growing Child in Wisconsin is 12 times higher than estimated by UNSCEAR for the Northern Latitude. (Range of monitored stations: 8.7-16 times higher.)

The average dose commitment to the Bone of a Growing Child in Wisconsin is 38 times higher than estimated for the Northern Latitude for internal Sr-90 and Cs-137. The range is from 28 times higher (Madison) to 51 times higher (Rice Lake).

The underestimation of dose commitment to the Wisconsin Adult group from internal Sr-90 and Cs-137 is less than to the Growing Child group by a factor of more than

TABLE XXVI

ACCUMULATED INTERNAL SR-90 PLUS CS-137 DOSE COMMITMENT
R^3 DIET, WISCONSIN / EXPECTED: NORTHERN LATITUDE (mrems)

	Accumulated Dose, Wisconsin		*Expected	Product of Ratio: Wisconsin/Expected	
	Average	Range		Average	Range
Total Body					
Growing Child	1055.452	784.077 – 1411.749	88	12	8.7 –16
Adult	491.89	355.59 – 646.02	88	5.6	4 – 7.3
Bone:					
Growing Child	4248.66	3157.7 – 5665.11	111	38.3	28.45–51.04
Adult	1896.56	1368.13 – 2501.063	111	17.09	12.3 –22.5

* Expected dose for Northern Latitude from United Nations Committee on the Effects of Atomic Radiation report in EPA, Radiological Quality of the Environment, United States, 1977, Table 4-9.

See Table II in Methodology herein.

2: Average monitored Wisconsin station underestimation by UNSCEAR by a factor of 5.6 for Total Body Dose Commitment; by a factor of 17.09 for Dose Commitment to Bone.

The Internal Sr-90 plus Cs-137 Dose Commitment for Wisconsin groups in Table XXVI were derived from records of milk monitoring (1963-1976) in Wisconsin. With the addition of Iodine-131 dose, these Wisconsin figures constitute the R3 Diet Dose Commitments.

Table XXVII compares Total Fission Dose commitment estimates for Wisconsin to those estimated by UNSCEAR for the Northern Latitude. For Wisconsin the Total Fission Dose Commitment was derived by adding other major internal doses and external doses to the basic R3 Diet dose in the ratio indicated by the UNSCEAR report. An exact quantification of Total Fission Dose cannot be derived for Wisconsin since no records for external dose or for internal dose from other nuclides are available for the state for the 1963-1976 period.

The factor of difference between Wisconsin dose estimates and the expected dose for the Total Fission commitment is somewhat less than for the comparison of only Internal Sr-90 and Cs-137 dose. The average factor of underestimation by UNSCEAR of Wisconsin dose: Total Body Total Fission Dose Commitment: 10.4 for Growing Child; 5 for Adult — Total Fission Dose Commitment to Bone: 32 for Growing Child; 14.4 for Adult.

The fact that the UNSCEAR underestimation of ingested radionuclides (Sr-90 and Cs-137), Table XXVI, is slightly more than the underestimation of the Total Fission dose (ingested plus external), Table XXVII, indicates that the methodology for deriving the estimated Total Fission Dose is conservative by a factor of about .15.

A comparison of Wisconsin dose to expected was also calculated for that part of the Dose Commitment which has already been delivered as accumulated Annual doses by 1977 (35% to 40% of the dose commitment). The Environmental Protection Agency has estimated that the Annual total fission dose to whole body accumulated by the United States population in the years 1963-1976 from

TABLE XXVII

TOTAL FISSION DOSE COMMITMENT
TOTAL FISSION DOSE, WISCONSIN/EXPECTED, NORTHERN LATITUDE (mrems)*

	Total Fission Dose Commitment, Wisconsin		Expected	Product of Ratio: Wisconsin/Expected	
	Average	Range		Average	Range
Total Body:					
Growing Child	2391.93	1777.39—3195.84	230	10.4	7.7—13.9
Adult	1147.2	831.06—1502.34	230	5.0	3.6— 6.5
Bone:					
Growing Child	8458	6315.2—11085.8	260	32	24.3—42.6
Adult	3742	2701.2— 4928.7	260	14.4	10.4—19

* Total Fission Dose Commitment, Wisconsin, is R3 Diet Dose (from ingested Sr-90, Cs-137 and I-131) plus additional internal dose from H-3, C-14 and Fe-55, and External doses in ratio to monitored R3 Dose as indicated by United Nations Scientific Committee on Effects of Atomic Radiation in EPA, Radiological Quality of the Environment in the United States, 1977, Table 4-9.

See Table II in Methodology herein.

bomb testing deposits and nuclear power is about 80 millirems. (67)

This accumulated Annual dose estimate is compared to the Wisconsin realistic dose in Table XXVIII. For the Wisconsin Growing Child the Realistic dose is from 8.3 to 15.2 times higher than expected for the U.S. population by EPA (average of monitoring stations: 11.2). For Adult the Realistic dose is from 4.1 to 7.4 times higher (average of 5.6) than expected.

The factor of underestimation is slightly higher for expected annual dose accumulation in the United States (Table XXVIII) than for expected dose commitment for the Northern Latitude (Table XXVII).

Realistic Fission Dose, Wisconsin, Compared to Natural Radiation or X-rays. The expected dose from nuclear fission is often quantified in terms of percentage of natural radiation or radiation due to medical X-rays. Natural radiation background in Wisconsin is estimated at 107 millirems/ year by the Environmental Protection Agency; X-ray mean active bone marrow dose is estimated at 103 millirems/ year by the EPA for the adult U.S. population. Table XXIX lists the annual per capita dose from various types of X-rays. The range is from .4 to 24.3 millirems/ year with only five types of X-rays estimated to deliver a dose of more than 5 millirems/year. Seventy-eight percent of the x-ray types listed by EPA deliver less than 5 millirems/year.

The natural radiation dose per year (approximately the same for X-rays) is compared in Table XXX to average annual whole body and bone dose to population groups in Wisconsin from nuclear fission.

First, annual dose from the R3 Diet (ingestion only of Sr-90, Cs-137 and I-131) to total body is compared. For the Growing Child the Fission dose was 26% of natural radiation dose (range: 20% to 36%). In 1976 (last year of dose period) the range was from 23% to 41% of natural radiation. For the Adult the Fission dose was 13% of natural radiation dose (range of 10% to 17%). For the Annual Dose in 1976 to Adults from the R3 Diet the Fission whole body dose was 14% of natural radiation.

For R3 Diet dose to Bone of Growing Children the yearly average was 10% more than natural radiation dose

TABLE XXVIII

ACCUMULATED ANNUAL FISSION DOSE TO TOTAL BODY (mrems) 1963-1976: WISCONSIN REALISTIC DOSE COMPARED TO U.S. EXPECTED

	Wisconsin Realistic Dose		Expected* U.S. Average	Product of Ratio: Wisconsin/Expected U.S.	
	Average	Range		Average	Range
**R^3 Diet Only:					
Growing Child	395.22	294.202-535.002	<80	>4.9	>3.7->6.7
Adult	197.95	143.54—261.54	<80	>2.5	>1.8->3.3
*** Total Fission Dose:					
Growing Child	896.0	666.9—1212.9	80	11.2	8.3-15.2
Adult	448.8	325.4— 592.95	80	5.6	4.1—7.4

* U.S. EPA estimate for average annual fission dose accumulation to U.S. population for 1963-1976 period. See Radiological Quality of the Environment in the United States, 1977, Table 4-10.

**Wisconsin R3 Diet Dose is from ingestion only of Sr-90, Cs-137 and I-131. The portion of the expected Annual dose accumulated by the average U.S. resident which is due to this pathway only is not specified by the U.S. EPA. The "expected" dose above is therefore indicated as "less than" the total dose from all pathways (<80 mrems), and the factors by which the Wisconsin Realistic Dose exceeds the "expected" U.S. Average Dose are indicated as "more than."

***Total Fission Dose is R3 Diet Dose plus additional internal dose and external dose. See Methodology.

TABLE XXIX

MEAN ACTIVE BONE MARROW DOSE
TO THE ADULT POPULATION (1970)

Examination	Mean active bone marrow dose per examination (mrad)	Annual per capita examination rate		Annual per capita dose (mrad ± S.E.)
Head and neck				
Skull	78	0.020		1.6±0.1
Cervical spine	52	0.022		1.2±0.2
Other	--	---		0.6±0.2
Thorax				
Chest-photofluoro.	44	0.073		3.2±0.3
Chest-radiographic	10	0.306		3.2±0.1
Thoracic spine	247	0.010		2.5±0.4
Ribs	143	0.009		1.3±0.2
Other	--	---		1.9±0.4
Upper abdomen				
Upper GI series (total)	535			24.3±4.7
Radiographic (subtotal)	294	0.046	13.5±4.3	
Fluoroscopic (subtotal)	241	0.045	10.8±1.9	
Scan	167			
Spot films	74			
Lumbar spine	347	0.023		8.1±0.8
Gall bladder (total)	168			3.7±0.4
Radiographic (subtotal)	129	0.027	3.5±0.3	
Fluoroscopic (subtotal)	39	0.006	0.2±0.3	
Scan	29			
Spot films	10			
Small bowel series	422	0.002		1.0±0.3
Other	--	---		2.1±1.0
Lower abdomen				
Barium enema (total)	875			21.2±1.8
Radiographic (subtotal)	497	0.024	11.9±1.0	
Fluoroscopic (subtotal)	378	0.024	9.3±1.5	
Scan	268			
Spot films	110			
IVP	420	0.024		10.1±0.6
Lumbosacral spine	450	0.013		5.7±0.7
Abdomen KUB	147	0.020		2.9±0.4
Other	--	---		0.4±0.2
Pelvis				
Pelvimetry	595	0.002		1.4±0.5
Pelvis	93	0.012		1.1±0.2
Hip	72	0.009		0.7±0.1
Other	--	---		1.2±0.7
Extremities				
Femur	21	0.002		0.04±0.02
Dental	9.4	0.312		2.9±0.2
			Total	103±5

From U.S. Environmental Protection Agency, Radiological Quality of the Environment in the United States, 1977, p. 213.

TABLE XXX

ANNUAL WISCONSIN FISSION DOSE, 1963-1976 COMPARED TO NATURAL RADIATION (millirems/year)

Fission: Accumulated Annual Dose, 1963-1976			*Natural	Product of Ratio: Natural/Fission	
Whole Body — R3 Diet **	Average per yr	Range		Average	Range
Growing Child:					
14-year average	28.23	21.01-38.21	107	.26	.20— .36
1976 Annual Dose	33.11	24.60-44.17	"	.31	.23— .41
Adult:					
14-year average	14.00	10.25-18.68	"	.13	.10— .17
1976 Annual Dose	14.77	10.64-19.46	"	.14	.10— .18
Bone Dose — R3 Diet **					
Growing Child	117	87.2-157.6	"	1.1	.81-1.5
Adult	50	35.9— 66.8	"	.47	.33— .62
Whole Body Total Fission Dose **					
Growing Child	64	47.6— 86.6	"	.60	.44— .81
Adult	32	23.2— 42.35	"	.30	.22— .40

* Natural Radiation as determined by U.S. EPA as average Wisconsin state dose in Natural Radiation Exposure in the United States, Report ORP/SID 72-1, June 1972. Applied by U.S. NRC in NUREG-0226, April, 1977, pp. 5-19 relating to Final Environmental Statement, Tyrone Energy Park.

** R3 Diet Dose is from ingestion only of three radionuclides: Sr-90, Cs-137 and I-131.

*** Total Fission Dose is the R3 Diet Dose plus internal radiation from three additional radionuclides plus external dose as determined by ratio of R3 to additions in UNSCEAR report of Dose Commitments, Northern Latitude and by Environmental Protection Agency. See Methodology.

(range of 19% less to 50% more). For Adults the annual Bone dose from the R3 diet averaged 47% of natural radiation (range of 33% to 62%).

The average annual whole body Total Fission Dose (R3 Diet plus external and other internal doses) was also compared to Natural Radiation in Table XXX. For the Growing Child the average Annual dose was 60% of natural radiation (range of 44% to 81%). For Adults the fission dose was from 22% to 40% of natural radiation.

Ordinarily it is expected that Annual dose from nuclear fission will be only 3% to 4% of natural radiation.

Figure VIII is a summary comparison of Annual fission dose findings in this study (Wisconsin average: 1963-1976) to expected doses to residents of the United States as outlined in U.S. Environmental Protection Agency, Radiological Quality of the Environment in the United States, 1977. Wisconsin Dose figures are rounded to nearest whole millirem. The estimated average Annual Total Fission Dose to Bone assumes only an average of 28% of the 50-year Dose Commitment and is meant to represent the bare minimum possible dose. (EPA studies do not estimate average Annual doses to bone from internal and external pathways.)

Reasons for Discrepancy between Realistic Fission Dose and Expected Dose

Some of the factors which account for these discrepancies between expected fission doses and the realistic doses calculated for Wisconsin population groups are apparent:

First: All expected doses are calculated for a population-weighted age distribution. In Wisconsin, 65% of the 1970 population was more than 17 years old — only 19% were age 1 to 10. Since the millirem/year dose from Sr-90 (which constitutes most of the bone dose) is much higher per picocurie of intake for children than for adults, a population-weighted Sr-90 dose estimate will always be an understatement of dose to the Growing Child. (68) This also explains the large differences in Growing Child and Adult doses in Wisconsin.

Conversely, Cs-137 will have more biological effect

FIGURE VIII:
ANNUAL FISSION RADIATION DOSE, WISCONSIN (1963-1976 average) COMPARED TO AVERAGE DOSE ESTIMATES

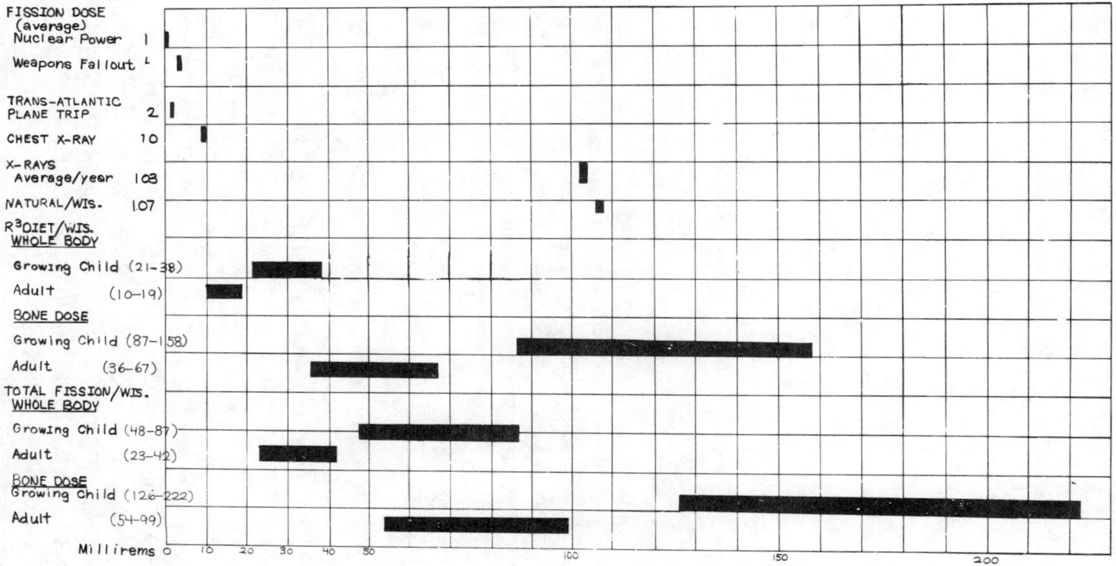

R^3 Diet Dose is from Sr-90, Cs-137, I-131 only in the diet.
Total Fission Dose includes external dose calculated from U. S. estimates proportioned to R^3 Diet.

per picocurie on Adults than on the Growing Child. A population-weighted dose estimate for Cs-137 would overestimate the Growing Child dose if Cesium dose exceeds Strontium dose. In Wisconsin the annual Adult dose from ingestion (R3 Diet) was higher from Cesium only in 1963 and 1964 when fallout was heaviest. For all other 12 years Strontium is by far the dominant radionuclide. This balance is expected to prevail for the U.S. as a whole and, therefore, after 1964 all population-weighted dose estimates from ingestion will understate dose to Growing Child.

Second: Strontium-90 is often not included in calculations of total body dose. The dose estimates for total body from Sr-90 to Wisconsin populations were derived from NRC Regulatory Guide 1.109. For total body dose to a Growing Child, Sr-90 yields 93.29 times as many millirems per picocurie ingested as does Cs-137. Most of the total body dose in Wisconsin is from ingested Sr-90. Sr-90 delivers a dose to total body, bone and GI-LLI according to Regulatory Guide 1.109. In other literature the Sr-90 dose is indicated for bone and bone marrow. Since bone marrow is a blood-forming organ, the dose is to the whole body. (See correspondence from Environmental Protection Agency making this equation. (69))

Any dose estimate which excludes Strontium-90 from contribution to whole body dose and calculates the dose almost exclusively from gamma radiation due to Cesium-137 will understate the actual total body dose by a very large factor.

Third: Whole body doses are sometimes presented only for the monitored radiation pathway to humans. Thus, since milk is the most frequently monitored pathway, only milk intake is considered in some estimates. In the R3 Diet, Dairy products account for only 34% of the Internal Dose from Sr-90 and 29% of the Internal Cs-137 Dose. In some estimates "Dairy" is equated with "fluid milk" intake overlooking the large contribution from cheese and other dairy products. (See Table 1 for ratios.) When doses from external radiation and other radionuclides are added to the Total Fission Dose, the fluid milk dose contribution becomes a small percentage of the whole. This practice also is misleading in that the public can be led to believe that radiation dose can be avoided by eliminating some foods, like milk, from the diet.

In some instances both the practice of including only Cesium-137, and Cesium-137 only in Dairy products, are combined to produce a gross underestimate of actual whole body dose. (See correspondence with Wisconsin Radiation Protection Section. (70))

Fourth: Estimates of expected dose are usually in terms of annual dose for a particular year accrued only from the radiation intake during that year. Only 4.4% of Sr-90 intake during any one year is delivered as an annual dose in that year — the other 95.6% of intake during that one year continues to deliver an annual dose to body cells every year for the next 50 years before the fission decay process is complete in the body. Every annual dose during those 50 years includes a percentage of the dose intake of that one year. The process of mounting accumulation of radiation dose years after the initial radiation dose is evident in Tables XXIII and XXIV and in Figures VI and VII.

The radiological impact of radionuclide intake in years past is a very real portion of any one year's dose. For a radionuclide such as Sr-90, with a long biological effective half-life, this could be the major portion of any one year's annual dose.

Fifth: Comparisons of fission dose to natural radiation may, in some cases, be underestimated by the preceding factors, but, in addition, the understatement relative to natural radiation is also due to the fact that only whole body dose is usually considered in comparisons to fission dose. The natural radiation dose to Bone, for instance, is the same as it is for whole body. (71) Since the dose to bone is much higher than whole body dose (from Sr-90) especially for Growing Children, the exclusion of bone dose from comparisons to "natural" dose overlooks a large "unnatural" impact. Thus, while the whole body average Annual R3 Diet dose is 26% less than natural radiation, the average annual Bone dose is 10% more than natural radiation. (See Table XXX.) Fission doses are all equally "unnatural" irrespective of particular organ impacted.

Finally: Most dose estimates are in terms of the annual dose which is delivered in any one year or accumulated over a number of years — to the exclusion of the dose commitment for the biological effective half-life of the radionuclide intake. This commitment is inevitable, from the moment of intake, according to the laws of physics and biology and is of considerable importance in predicting future doses.

The following section explores the possibility that the wide discrepancy between the Wisconsin Fission dose and the Expected dose could be due to an unusually high radiation exposure in the State.

Historical Trend: Wisconsin Fission Dose Compared to United States

Comparison of the nationwide trend in levels of Sr-90 and Cs-137 in monitored milk (1963-1976) to levels at the six monitoring stations in Wisconsin is shown in Figure IX. Although gaps in monitoring at Wisconsin stations and imprecision of measuresments (as discussed in Tables V-IX) impose limits on this comparison, some conclusions can be drawn.

For Cs-137, a comparison of national to state trends (Figure IX) is limited by the increasing difficulty of analytical precision as concentrations decrease. Table VIII has demonstrated that for Cs-137 concentrations during the first three years (1963-1965), analytical precision by the State laboratory was very good (within 0.1 to 6.1% of EPA cross-check analysis). The contamination at state monitoring stations during these years is comparable to national levels with the state extremes straddling the national average line in Figure IX. Northern, Northwest and North-central stations are consistently well above the state average -- Southern, Southwestern and Northeastern stations are below. (See Table XI for pCi/l concentrations.) The divergence from a mean decreases noticeably during the early period indicating less recent fallout and leveling out of contamination among stations.

From 1966 through 1968 the state extremes suddenly drop below national average. Table VIII indicates that the state monitoring during this period may have underestimated concentrations by 16.5% to 73.1% as analytical pre-

FIGURE IX

CS-137 AND SR-90 (pCi/1) IN MONITORED MILK:
U.S. AND WISCONSIN (1963-1976)

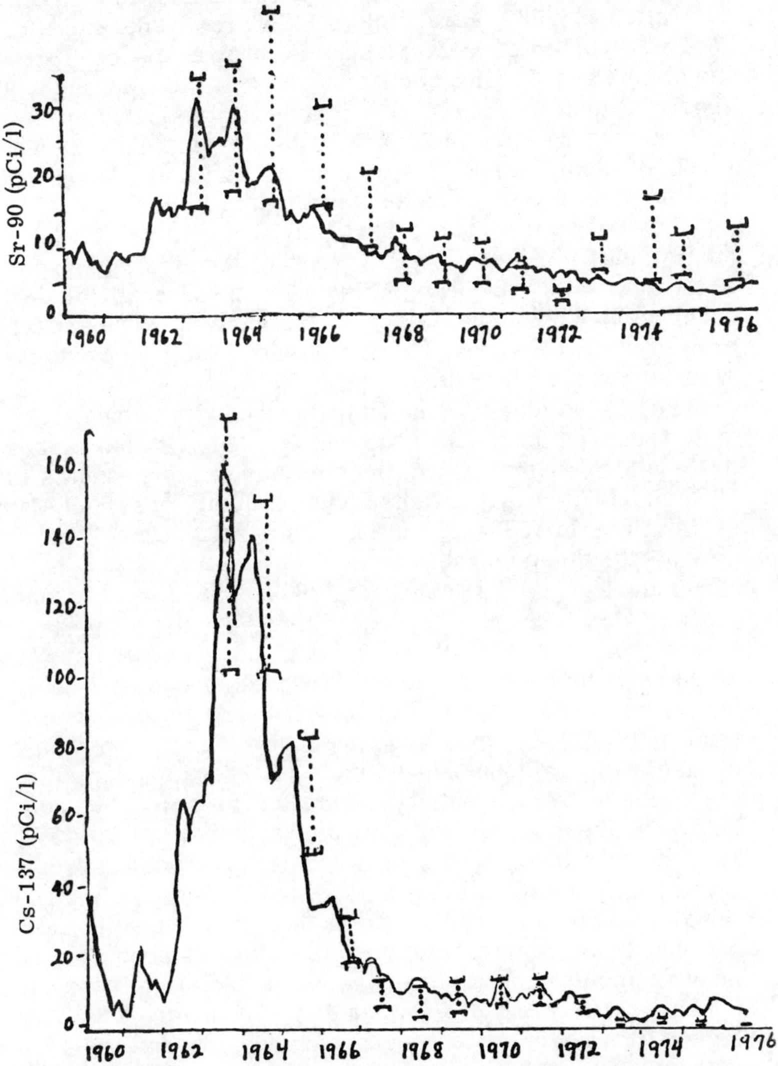

Solid Line: U.S. monthly average concentrations from U.S. EPA
Radiological Quality of the Environment in the U.S., 1977.
Vertical Broken Lines: Range of annual average concentrations at Wisconsin milk-monitoring stations. See Tables X and XI.

cision drops in proportion to radionuclide concentration.

In the last four years (1973-1976) all state stations are well below national averages. For these years Table VIII indicates that state analyses may underestimate concentrations by 92% according to EPA cross-check of samples at reported concentrations. It is unreasonable that the state Cs-137 fallout deposits of 1963-1964, at and well above national averages, could have disappeared at a much faster rate than national fallout levels declined. The levels of Strontium-90 (depicted also on Figure IX) do not decline at a rate much faster than national levels.

The only valid conclusion, therefore, from Cs-137 monitoring relative to a state/national trend comparison is that in the North — Northwest — North-central tier of counties in Wisconsin, fallout in the early 60's was heavier than nationally while the Southwest — South — Northeast area fallout was at or below national levels.

For Strontium-90, the limiting condition for national/state trend comparison is the omission of milk monitoring in the state particularly in the summer grazing months in 1963 and 1972 at all stations and at the LaCrosse and Madison stations in 1973 and the LaCrosse station in 1975. (See discussion and Table IX.)

Figure IX shows that from 1963 to 1965 Sr-90 levels in the state increased steadily while national average levels decreased steadily. This contrast may not be actual since state monitoring was not initiated until August and September 1963; the monitoring frequency improved somewhat in 1964-1965 which may account for an apparent increase in levels. The state levels in 1963-1964 straddle the national average with the same divergence above and below the mean among stations as was noted for Cs-137. In 1965 state levels are well in excess of national Sr-90 levels, and in 1966 all Wisconsin stations are above national average. Interpretation of this change relative to national levels is inconclusive due to lack of regular monthly monitoring in these years. Two possibilities are: (1) that state levels of Sr-90 in 1963-1964 were actually higher than sparse monitoring indicates and that the 1965 extremes are more representative of early fallout levels; or (2) that Wisconsin had heavy fresh fallout in 1965 which did not affect the nation as a whole. The progressive decline from 1963 to 1965 for Cs-137, following the decline slope of national average, indicates that the first possibility is the more probable of the two. Cs-137 is a better indicator of fresh fallout since it is not taken up from past ground deposits as readily as Sr-90.

In 1967 no Sr-90 monitoring was done by the state except for two January tests — the level charted on Figure IX has been determined by an average of 1966 and 1968 levels assuming a gradual decrease in levels during this period which was characteristic of the nation as a whole.

In the 1968-1970 period Wisconsin stations straddle the national average for Sr-90. The divergence among stations from a mean contamination level decreases steadily (as for Cs-137) indicating a leveling out of previous fallout deposits with decreasing fresh fallout. Monitoring is fairly consistent during these years when it appears that effects from early Sr-90 fallout on Wisconsin had diminished to a level about equal to the national average levels.

In 1971, and more dramatically, in 1972, Sr-90 levels in Wisconsin suddenly drop below national levels. The realism of this sudden drop from levels which had previously been consistently at or above national levels is questionable. A lack of monitoring may be responsible: In 1971 apparently only one Sr-90 test was analyzed for the September through December period at 5 stations since the concentration levels do not change at all. At the 6th station, La Crosse (Genoa), tests for each of these months show a range by a factor of 4. The probability of the same actual factor of range, September to December, at the other 5 stations is evident.

In 1972 the state dropped Sr-90 monitoring for most of the summer grazing months throughout the state. Thus, particularly for 1972, the drop below national average is highly questionable.

From 1973 through 1976 all Wisconsin stations are above the national average. Monitoring during this period is more frequent and regular than in any earlier period with the exception of the La Crosse (Genoa) station where 5 summer months are without monitoring (1973 and 1975) and Madison with 2 summer months missing in 1973. The divergence from a mean among state stations widens suggesting recent uneven deposition of Sr-90 fallout. This is particularly true for 1974 when a factor of about 3 differentiates highest from lowest concentrations at monitoring stations.

In conclusion, Sr-90 milk monitoring relative to state/national trends indicates that Wisconsin Sr-90 contamination was greater than the national average in the early 1960's. In the 1968 to 1970 period the early fallout had evened out among stations considerably and levels were comparable to national averages. The 1972 data is unreliable for comparative purposes due to lack of summer monitoring at all stations (to a lesser extent, 1971 data is also inconclusive).

In the 1973-1976 period Wisconsin levels of Sr-90 increase to levels well above national levels while the average U.S. levels continue to decline slowly.

Figure X compares the cumulative deposition of Sr-90 in the Northern Hemisphere during this period to the state annual average pCi/l Sr-90 in milk. Cumulative soil deposits drop gradually from 1966 through 1976. The U.S. average concentrations of Sr-90 in milk follows the gradual deposits in soil – from 1970 through 1976 these two slopes are almost identical. However, Sr-90 concentrations in Wisconsin milk plunge below the line sharply in 1972 (year of especially sparse monitoring) and in 1973 through 1975 Wisconsin milk concentrations suddenly soar well above the slopes of cumulative soil deposits and national average milk concentrations. The level of Sr-90 contamination of milk in Wisconsin in 1974 is up to pre-1968 levels. The demarcated area on Figure X above cumulative U.S. Sr-90 deposits indicates the additional deposits which would be necessary to account for the 1973-1975 Wisconsin milk concentrations. At the bottom part of Figure X the demarcated area indicates the excess Sr-90 concentration in Wisconsin milk compared to national average. It is very unlikely that the 1973-1976 rise in Wisconsin Sr-90 is due to heavier-than-usual uptake from previous cumulative deposits. As R. Scott Russell explains:

> Extensive investigations of the contamination of milk with strontium-90 from worldwide fallout have in recent years provided clear evidence that in times

FIGURE X

STRONTIUM-90: CUMULATIVE DEPOSITS, NORTHERN HEMISPHERE
PCI/L IN MILK; WISCONSIN, ANNUAL AVERAGE AND
UNITED STATES, ANNUAL AVERAGE

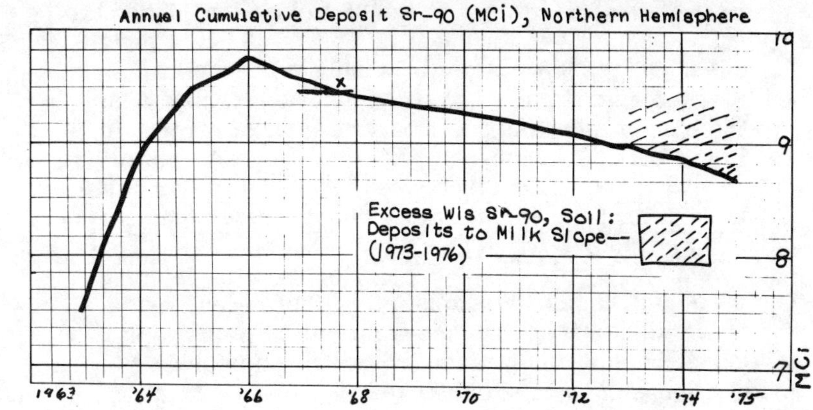

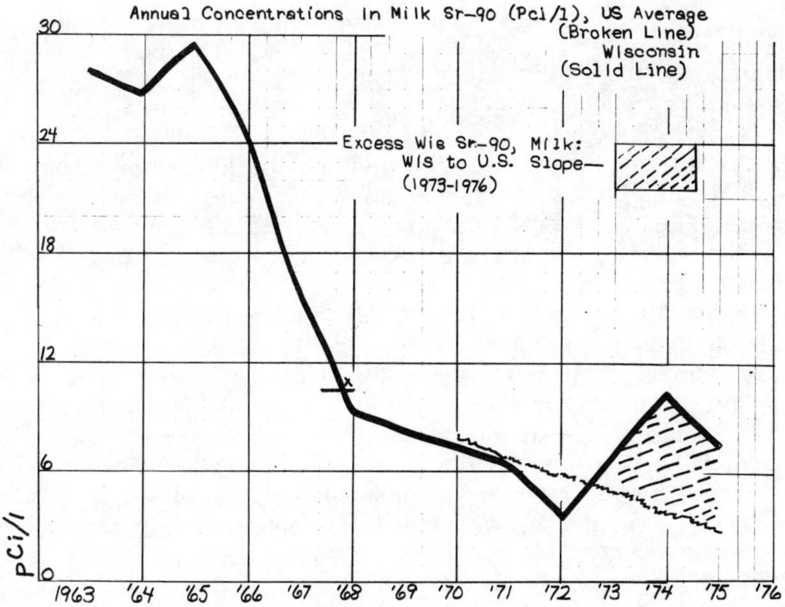

X indicates pre-1968 point comparable to 1974 peak of Sr-90 concentration in Wisconsin milk (lower graph) and soil deposit of Sr-90 in the U. S. at the same date (upper graph).

References: U. S. EPA, Radiological Quality of the Environment in the U. S., 1977, and milk monitoring by Radiation Protection Section, Dept. of Health and Social Services, Wisconsin.

of appreciable deposition the direct contamination of pastures with recently deposited material is a major source of radioactivity in the diet of cattle. When the suspension of weapons testing caused the rate of fallout to decrease considerably in 1959, there was a marked reduction in the contamination of milk despite the slowly increasing cumulative total of strontium-90 in the soil. Subsequently, when testing was resumed and the rate of fallout increased, the levels of strontium-90 in milk rose considerably more steeply than the cumulative deposit. (72) (Emphasis added.)

The steep increase in Wisconsin Sr-90 in recent years clearly indicates fresh radiation fallout unevenly distributed among the state monitoring regions. Possible sources of radiation in Wisconsin are explored in the following section.

Sources of Fission Dose in Wisconsin

A thorough examination of possible sources of the nuclear fission dose to the Wisconsin population is beyond the scope of this radiation dose study. However, some preliminary deductions emerge when the most obviously possible sources are analyzed in relation to the six monitoring stations.

Table XXXI was prepared as a summary of the most probable known sources of fission dose in addition to aerial bomb-testing finally banned in 1963. It is estimated that uptake of early bomb-testing fallout from soil deposits has been decreasing at the rate of 13 to 14% annually. (73) Recent fallout is estimated to contribute from 80 to 97% of Sr-90 dose and 70% of Cs-137 dose. (74) (Rises in Sr-90 levels as illustrated by Wisconsin milk monitoring in very recent years must be attributable to recent fallout.) Table XXXI shows that, with the exception of the Dresden I (Ill.) nuclear reactor, Wisconsin was not close to any direct source releasing new fallout radionuclides until late 1969. (75) From late 1969 through 1974, 14 nuclear power plants began operation in Wisconsin or adjacent states. The People's Republic of China exploded 8 nuclear bombs and underground tests in Nevada were reported to have vented in a northeasterly direction. None of the bomb-testing events caused noticeable nationwide rises of Sr-90. National averages have continued to decline.

In addition to these sources of fission doses there are innumerable military weapons facilities (noteably at Hanford, Washington to the west), experimental federal facilities (such as the Idaho reactor testing complex to the west) and federal and state research reactors and other fission facilities.

In the early period some fresh fallout is indicated in 1965-1966 by rises in I-131 levels (See Table XII). The Northeastern station (Green Bay) is the least affected; the Northwestern region (Rice Lake) is affected only by the 1965 rise. Delayed fallout from the Chinese bomb test, 5/14/1965, underground testing in Nevada on 6/14/1965 or the Dresden plants in Illinois appear to be the most probable sources. In 1966 the South—Southwest—North-central area have the highest rise. Table XXXI indicates the possibility of fallout from venting of underground tests in Nevada, 6/15/1966, or gaseous releases from the Dresden reactor downwind to the south in summer months, or delayed fallout from a Chinese test of 5/9/1966.

The least explainable statewide rise in I-131 occurs in May 1970. (Cs-137 levels also rise in the following month — See Tables XI and XII.) All stations rise to pCi/l levels of I-131 of 30 to 40 (except Eau Claire with slightly less rise). Such levels statewide are highly unusual. The only comparably high levels across the state occurred in June 1965 and June 1966. The pattern of the 1970 fallout differs from 1965-1966 fallout in that Eau Claire and Rice Lake, the stations with highest levels from earlier fallout, are least affected by this 1970 episode. It does not appear to have deposited heavily in the North — Northwest area but on the South—Southwest — Central and Northeast area.

There are no Chinese bomb tests in this 1970 period of high I-131 levels. Possible sources are an underground Nevada test (Mint Leaf) on 5/5/1970 which developed leakage on the following day carried by winds "to the east", the Dresden reactors or the La Crosse plant.

The three cited episodes are the only instances resulting in statewide rises in I-131 to the 30 to 40 pCi/l level of concentration.

TABLE XXXI

FISSION CHRONOLOGY – WISCONSIN AREA, 1963-1976

Operating Nuclear Plants – Start-up dates and gaseous releases (curies)

Dresden, I, II, III – start-up 1959, 1/1970, 1/1971 (Ill.)

Year	Releases	Year	Releases
1962	284,000	1971	1,333,000
1964	521,000	1972	1,306,000
1965	610,000	1973	1,720,000
1966	736,000	1974	725,000
1969	862,000	1975	880,000
1970	1,160.000	1976	484,300

Genoa (LaCrosse) -- start-up 7/1969 (Wis.)

1972	31,000
1973	91,000
1974	49,000
1975	57,000
1976	124,000

Point Beach I, II – start-up 11/1970, 5/1972 (Wis.)

1972	3
1973	5,750
1974	9,740
1975	44,500
1976	1,910

Monticello – start-up 12/1970 (Minn.)

1971	76,000	1974	1,480,000
1972	751,000	1975	155,000
1973	810,000	1976	11,400

Quad Cities I, II – start-up 10/1971, 4/1972 (Ill.)

1972	132,000
1973	900,000
1974	950,000
1975	110,000
1976	36,000

Zion I, II — start-up 1973, 1974 (Ill.)

Prairie Island I, II, start-up 1973, 1974 (Minn.)

Kewaunee, start-up 1974 (Wis.)

1974	3,150
1975	2,450
1976	1,400

Duane Arnold, start-up 1974 (Iowa)

1975	1,580
1976	5,260

Chinese Aerial Bomb Testing Dates:

10/16/64	6/17/68	1/ 7/72
5/14/65	12/27/68	3/18/72
5/ 9/66	9/22/69 (underground)	6/26/73
10/27/66	10/11/70	6/17/74
12/28/66	11/13/71	9/26/76
		11/17/76

Underground Tests in Nevada Reported as Venting with Cloud Travelling in Northeast Direction

12/16/64	Parrot Test	250,000 curies
6/16/65	Diluted Water	5,900 curies
3/ 5/66	Red Hot	1,000,000 curies
4/25/66	Pin Stripe	350,000 curies
6/15/66	Double-Play	750,000 curies
9/12/66	Derringer	12,000 curies
8/31/67	Door Mist	400,000 curies
4/21/67	Snubber	54,000 curies
5/ 5/70	Mint Leaf (leakage on 5/6/70)	395,000 curies
12/18/70	Baneberry	3,000,000 curies

References:

Carter and Moghissi, "Three Decades of Nuclear Testing," Health Physics, July, 1977.

U.S. Nuclear Regulatory Commission, NUREG-0367.

Eisenbud, Environmental Radioactivity, 1973.

U.S. Atomic Energy Commission, Division of Occupational Safety, Summary Information on Accidental Releases of Radioactivity to the Atmosphere from Underground Nuclear Detonations Designed for Containment, August 5, 1963-June 30, 1971. June, 1971.

Since fresh fallout is indicated more quickly by I-131 milk concentrations, a more detailed examination of levels for the 1973-1976 period was prepared to determine whether or not the excess fallout in Wisconsin could have resulted from Chinese bomb tests in the years 1973, 1974, and 1976.

Table XXXII lists Iodine-131 concentrations above 5 pCi/l during these years by date as well as dates of the bomb tests. Airborne radioactive materials would be expected to arrive over the North American Continent within 4 to 7 days following tests. (76) Of 90 listed I-131 levels over 5 pCi/l in the state during these years only one is clearly a possible result of Chinese bomb tests (7/02/1973 at Rice Lake). One is a slight possibility of delayed effect (7/16/1973 at Eau Claire). Elevated I-131 tests at Green Bay, Madison and La Crosse in November 1976 may have been enhanced by Chinese testing fallout, but, since elevations began before the test (11/02 and 11/05/1976), they cannot have been initiated by the Chinese test of 11/17/1976.

It must be concluded that fallout in Wisconsin during the 1973-1976 period was only slightly if at all a product of nuclear weapons testing by the People's Republic of China. It is beyond the scope of this study to definitively trace sources of the radioactive materials. The proximity of these state monitoring stations to particular nuclear facilities may suggest sources. (See Figure I.) But in this 1973 to 1976 period Wisconsin is under the influence of 15 midwest nuclear plants. Wind and rain and snow carry radioactivity over many miles. Accurate logs of daily nuclear plant releases coordinated with frequent, accurate testing for radionuclides in foods, and wind and weather information are difficult or impossible to obtain. However, a closer look at stations in the state to determine which areas have been affected most by recent fallout or earlier fallout was undertaken to discern trends.

It was observed that the order of radioactive pollution among stations was more constant during the first half of the 14-year period under study than the second half. There was considerably more switching of rank in pollution status in the latter period. In order to determine if any substantial changes had taken place in relative radiation pollution status from the first to second half of the monitoring period, rank numbers from 1 to 6 were assigned for each station — number 1 indicating highest pollution status. A comparison of the sum of rank numbers for each station for both periods yielded the ratios charted in Figure XI (sum of rank numbers, 1st half/Sum of rank numbers, 2nd half). The highest ratio product, 1.4, at Wausau indicates this station improved its relative pollution rank in the state more than any other station. Rice Lake (ratio product, 1.3) also improved. However, Eau Claire showed the most decline in pollution status (ratio product, 0.63) in the second half of the period. These three stations have higher final dose accumulations than the other three stations but, obviously, one is receiving even higher levels of radiation in recent years (Eau Claire) while the other two have somewhat improved during the more recent period. The other stations do not show as much change from first to second half with La Crosse and Green Bay showing practically no change and Madison declinding in pollution status somewhat in the 2nd half.

Since it is particularly in the last four years that the Sr-90 levels in the state all suddenly rose above national average levels, these years were analyzed in more detail. Table XXXIII indicates the degree to which each of the stations exceeded the national average in the 1973-1976 period. The average Sr-90 concentrations during this period in Wisconsin exceeded the national levels from 153% to 263%. In 1971 the statewide average was slightly below the national average. Using 1971 as a base year, the excess Sr-90 (1973-1976 average) over national levels was computed for each station. Eau Claire and Green Bay show the highest excess: 9.39 pCi/l and 6.215 pCi/l respectively. Following in order of greatest excess are Rice Lake at 4.385 pCi/l; Wausau at 3.58 pCi/l; La Crosse by 2.64 pCi/l and Madison at 2.225 pCi/l.

All stations were subject to increases in radiation in the 1973-1976 period over the national average decline. With 1971 as a base year, the national average Sr-90 concentration declined from 7 pCi/l to 4 pCi/l while the state average increased from 6.5 pCi/l to 8.44 pCi/l.

However, from year to year, there are very noticeable differences among stations in rate of change. From

TABLE XXXII

WISCONSIN IODINE-131 TEST RESULTS ABOVE 5 pCi/l AND CHINESE BOMB TESTS, 1973, 1974, 1975

Date	Rice Lake	Eau Claire	Wausau
1973	1/19		1/11
	2/12		2/21
			2/27
			3/06
		4/06	
Chinese Test – 6/26		5/21	
	7/02		
	7/27	7/16	7/27
	8/08	8/08	8/09
		9/04	9/18
		10/02	
1974	1/04		
	1/07		
	1/28	1/21	
	*2/07	2/04	*2/09
	2/22	2/20	
	4/10		
Chinese Test – 6/07	5/02		
	7/15		
	11/29		
1976		1/16	
			3/23
		5/05	5/04
		6/04	6/02
		7/14	
Chinese Test – 9/26		8/26	
			10/19
Chinese Test – 11/17/'76			
	11/17		
		12/02	
		12/15	12/14

* Iodine-131 concentrations in milk above 10 pCi/l.

** A radiochemical analysis of 25.66 pCi/l "was reported as 'lost.'" Reference: Radiological Monitoring Study in progress by Another Mother Fund for Peace in reference to Memo dated 7/12/76 from Wisconsin Laboratory to Radiation Protection Section Chief.

	Green Bay	Madison	LaCrosse
1973	1/26	1/26	
		2/02	
	3/07		
	5/25	5/18	
		6/05	
			7/02
		7/27	7/05
	8/24		
	9/18		
	11/15		
		1/04	
1974	*1/28		
	*2/07	*2/08	
		2/21	
	9/09	7/27	
	10/08		10/08
1976	1/21		
	3/10	3/23	3/09
		4/07	
	5/06	5/06	
		6/15	**6/1/'76
	7/08		
	8/11		
	9/23		
	10/06		
	10/20		
	11/02	11/03	11/05
	11/17	11/15	
		11/29	
		11/30	11/30
	12/14	12/14	
	12/29		

FIGURE XI

IMPROVEMENT OF RELATIVE RADIOACTIVE POLLUTION RANK
AMONG MONITORED STATIONS: 1970-1976 period over 1963-1969 period

RATIOS: 1--No Change
>1--Improving Status
<1--Declining Status

TABLE XXXIII

COMPARISON OF NATIONAL AND WISCONSIN Sr-90 pCi/l IN MILK -- 1971, 1973-1976

	Rice Lake	Eau Claire	Wausau	La Crosse	Green Bay	Madison	State Average	U.S. Average	Excess Wis./U.S.
1971 (a)	8.71	6.06	7.56	6.74	4.3	5.7	6.5	7	0
1973	9.83	11.0	6.58	5.46	6.38	6.7	7.66	5	153%
1974	10.3	16.6	10.90	5.22	11.32	7.1	10.24	4	256%
1975	11.00	10.7	6.66	7.5	6.32	5.4	7.9	3	263%
1976	9.53	11.5	8.42	7.33	6.04	4.9	7.95	4	199%
1973-1976 Average (b)	10.165	12.45	8.14	6.38	7.515	4.525	8.44	4	211%
Change pCi/l (a) to (b):	+1.385	+6.39	+.58	-.36	+3.215	-.775	+1.94	-3	
Excess pCi/l – U.S. Average/Wisconsin stations: 1971 to 1973-1976 Average:	4.385	9.39	3.58	2.64	6.215	2.225	4.94	—	

1971 to 1973 all stations except La Crosse and Wausau go up (a smaller decrease at Wausau). Eau Claire experiences by far the greatest increase; next is Green Bay. In 1974 all stations except La Crosse increase with most measurable increases at Eau Claire, Green Bay and Wausaw. In 1975 while 4 stations decrease markedly, Rice Lake rises slightly and La Crosse has an increase of more than 50%. In 1976 the state average increases only slightly over 1975 with increases only at Eau Claire and Wausau.

The station least likely to follow state trends is La-Crosse. In this recent period (see previous discussion) this station has had less frequent Sr-90 monitoring than any other station. It is also the only station for which computer printouts of monitoring data were unavailable (discontinued in June 1972) and the only station for which an Annual Environmental report for a nuclear station (Genoa) were turned to for data. An analysis of rises of more than 100% in monthly Sr-90 concentrations over the previous month was made to determine how many of such increases were not noted at any other state station in the same month. In 1973, 40% of such rises in the state occurred at La-Crosse; in 1975 50% occurred at La Crosse. Sources of these unexplained rises, not typical of the rest of the state, could have been the Genoa plant or plants to the west in Minnesota. Or the untypical fluctuations in radiation in this area may be a reflection of the difference in the nature of the basic monitoring data. This possibility is further illustrated in Table XXXII, reviewing Iodine-131 concentrations above 5 pCi/l for three recent years which reveals that, while from 13 to 22 dates were recorded for the 5 other stations, only 6 were recorded for La Crosse. The realism of such relative freedom from I-131 pollution in these recent years is doubtful when in 1971 (before computer printouts were dropped) the La Crosse area had the 2nd highest I-131 levels in the state (Eau Claire only .06 pCi/l higher average than La Crosse — see Table XII).

The Eau Claire station shows both the greatest deterioration of pollution status (Figure XI) and the greatest degree of excess Sr-90 in the 1973-1976 period over national average (Table XXIII). In the early 1960's the Wausau and Eau Claire regions (North-central and West-central) appear to have been about equally affected by fallout. However, Figure XI shows Wausau (geographically furthest removed from an obvious fission source among stations) with the greatest improvement in pollution status, while Eau Claire (in closer proximity to Minnesota reactors in the recent years) shows the greatest deterioration in pollution status. The plants to the south may also influence Eau Claire to a greater extent than Wausau. In 1974, when Eau Claire experienced the greatest degree of increase in Sr-90 over a previous year, the Monticello reactor to the west released more radioactivity than any other plant in the nation. This station is close to a proposed new nuclear reactor, the Tyrone Energy Park. The evidence of recent trends indicates that the construction and operation of another reactor in close proximity to this station will result in further radioactive pollution. (Two of the three highest recorded Sr-90 levels in the 14-year period at Eau Claire occurred in June and August 1974 – 39.2 pCi/l and 41 pCi/l respectively; 42 pCi/l in 1963.)

Green Bay (Northeast) is located northwest of the Point Beach – Kewaunee plants on Lake Michigan which began operation in 1970, 1972 and 1974. Some of Green Bay's water supply comes from near Kewaunee in Lake Michigan. This station did not improve in pollution rank status between the 1st and 2nd half of the 14-year period. Table XXXIII comparing 1971 to 1973-1976 radiation levels indicates that radiation increases over national average in this most recent 4-year period were second only to Eau Claire. The highest recorded level of Sr-90 at Green Bay occurred in July 1974 (35 pCi/l; next highest, 29 pCi/l in 1963). And in 1974 the Point Beach plant exceeded predicted exposures from gaseous releases by a factor of 2. (77) The release record of the plant (see Table XXXI) shows increases as the plant developed problems in 1974-1975 over its earlier low gaseous releases. Radiation pollution in this area may increase or decrease depending upon plant operation. A new nuclear plant is planned to the south of this station at Haven on Lake Michigan.

The Madison area (South–Southeast) has the lowest accumulated radiation dose. Figure XI shows a slight deterioration in the last half of the monitoring period in relative pollution status. This station is not geographically close to a nuclear facility but it is located in a ring of plants to the south and west. In addition, as a major research University site and the state capitol, the city itself contains fission sources. Although Madison shows the least increase in radiation in the 1973-1976 period over 1971 among stations, it is still well above national average and its relative pollution rank may continue to rise as more plants are in construction and proposed stages directly to the south. (See Figure II.)

Wausau (North-central) is further geographically from a direct fission source than any in the state and shows the most improvement in pollution rank in the 2nd half of the monitoring period. This station was subject to more than average fallout in the early period. Cesium-137 is particularly noticeable. The 14-year accumulated dose is above state average. There is an indication that this area is affected by the same fallout as those to the west and northwest but to a lesser degree. It is subject to winds from all directions at different times of the year, and, therefore, may reflect some fallout from any of the sources. Indications are that this area may continue to improve in relative pollution rank as effects of early bomb-testing decline. However, it will continue to be affected by fallout particularly from the west from operating plants. The proposed Tyrone plant may result in heavier fallout. If the proposed Rudolph plant is constructed and operated in this region, Wausau may not be able to improve pollution rank. Presently it is well above national levels of Sr-90 contamination.

The Rice Lake station (Northwest) has the highest 14-year accumulated dose of all stations. In 1965 it was far above all stations in Sr-90 average (45 pCi/l compared to a state average of 29). Levels declined steadily until 1973-1976. Increases in 1973 and 1974 are parallel to those of Eau Claire, directly south, but to a lesser degree, indicating that the same new influence may have affected both of these areas in the later period. The monitoring station in this area does not appear to cover the extreme northern portions. This should be examined further since early monitoring in Minnesota shows much higher levels in Duluth (Minn.), adjacent to Superior (Wis.) than the Rice Lake records indicate. Cesium-137 concentrations run to over 400 pCi/l at Duluth in the 1963-1965 period while highest Rice Lake concentrations are in the 200 pCi/l range with only one recorded level over 300 pCi/l. Indications are that the high accumulated dose at Rice Lake revealed by Wisconsin monitoring may underestimate the accumulated dose in an area north and east of Rice Lake. (78) If radiation releases from the Monticello plant to the west continue and if the proposed Tyrone plant to the south is constructed and operated, this hardest-hit northwest region will probably not have an opportunity to improve its pollution rank status. Without additional new fallout in the recent period the heavy early bomb-testing effects could have diminished appreciably by now.

The prospect for reduction of radiation pollution in the future in Wisconsin will depend upon the condition of operating plants and decisions concerning new proposed plants. Doses to all individuals, however, will continue annually from biological impact annually of doses already committed from past intake into body cells. To date, only 35 to 40% of the accumulated commitment within body tissue and bone has decayed. Sixty to sixty-five percent of the dose is yet to be delivered over the next 50 or more years.

Expected Health Effects of Radiation Dose in Wisconsin

Various estimates have been made in the scientific community concerning the increased risk of cancers relative to accumulated radiation dose. These estimates have been revised continually as new studies are completed. The most recent major study of increased risks per radiation accumulated is that of cancer deaths among nuclear workers at the Hanford (Washington) facility. (79) Increased risk of cancers per millirem of accumulated radiation were found to be much higher than in previous estimates. One of the reasons cited for the new findings was that the Hanford population, unlike the survivors of the atomic bombing of Japan, represent a typical population, not subject to death and disease from many atypical causes as atomic bomb survivors. Subjects are therefore more likely to die of radiation-initiated cancers in a typican population period than in a violent, war-ridden population, where many other factors may cause death before cancer latent periods are completed.

Variations in degree of risk relative to age at time of irradiation have concluded that the risk is much higher to younger individuals. Table XXXIV reviews findings of variation in cancer induction per rad with age. The risk to individuals under age 6 is 10 times higher than to those in the 31-40 age bracket.

TABLE XXXIV

VARIATION IN CANCER INDUCTION PER RAD WITH AGE

Age at Irradiation (years)	Increase in Cancer Mortality per Rad (in the Plateau Region) (per cent)
In utero	50
0-5	10
6-10	8
11-15	6
16-20	4
21-30	2
31-40	1
41-50	0.5
51-60	0.25
61 and beyond	Assumed negligible

From: John W. Gofman, "The Cancer and Leukemia Consequences of Medical X-Rays," Table 1, Osteopathic Annals, November 1975.

A mean age at irradiation was calculated for the Wisconsin Growing Child population, studied in this dose estimate case, on the basis of annual dose each year 1963-1976. The age-specific increase cancer risk factor for the Growing Child population was calculated as 7.2 for Annual doses delivered to whole body (1963-1976). It was assumed that the age at irradiation for the Hanford workers was between 21 and 40, yielding an increase cancer risk factor of 1.5. The ratio of increased risk: (Hanford workers/Wisconsin Growing Child) was 1/5 (rounded figure). It was, therefore, concluded that the radiation dose level expected to double the risk of various cancers as predicted from Hanford workers (doubling-dose) would be less by 1/5 for the Growing Child in the Wisconsin case study.

Table XXXV was prepared to show the expected age-adjusted doubling-dose for cancer risk for the Growing Child and the percent increased risk from accumulated Annual whole body dose through 1977. (A similar increased risk estimate was not prepared for the Adult group since the Annual dose record does not include doses delivered from pre-1963 commitments.)

Estimates are for both the R3 Diet Annual Dose Accumulation (Sr-90, Cs-137 and I-131 in food only): and for the Total Fission Dose Accumulations (other internal doses and external dose added).

The current increased risk of all cancers to males from R3 Diet Dose only is 6% (range of 4 to 8%) – to females: 23% (range of 17 to 31%). From Total Fission Dose increased risk of all cancers to males is 13% (range of 10 to 18%) – to females: 51% (range of 38 to 70%).

The increased risk for all High Sensitivity cancers (those most likely to be induced by radiation according to the ICRP) from R3 Diet Dose only is 14% (range of 11 to 20%) – from Total Fission Dose: 32% (range of 22 to 44%).

The increased risk of Myeloma and Myeloid Leukemia from R3 Diet Dose only is 55% (range of 41 to 74%). From Total Fission Dose the increased risk is 124% (range of 93 to 168%).

The larger increased cancer risk for females is based on a much smaller number of cases in the Hanford employee study than risk for males and is, therefore, less certain.

Dr. John W. Gofman's estimate of a doubling-dose of 43.5 rems for all cancers was also applied to accumulated annual doses for the Wisconsin Growing Child group. Adjusted for age at irradiation, the Gofman doubling-dose yields from 3.38% to 6.15% increased cancer risk from the accumulated Annual R3 Diet Dose (a state average increased risk of 4.54%); and, 7.7% to 13.94% increased

TABLE XXXV

CURRENT RISK OF CANCERS TO GROWING CHILD, WISCONSIN FROM ACCUMULATED TOTAL BODY ANNUAL FISSION DOSE 1963-1976

Wisconsin Dose (mrem)	*Doubling dose (mrem)		Percentage Increased Risk
R3 Diet: 395.22 (294.2 - 535)	Expected	Age-adjusted	
Myeloma & Myeloid Leukemia	3,600	720	55 (41- 74)
**All High Sensitivity Cancers	13,900	2,780	14 (11- 20)
All Cancers (males)	33,700	6,740	6 (4- 8)
All Cancers (females)	8,700	1,740	23 (17- 31)
Total Fission: 896 (666.9-1212.9)			
Myeloma & Myeloid Leukemia	3,600	720	124 (93-168)
**All High Sensitivity Cancers	13,900	2,780	32 (24- 44)
All Cancers (males)	33,700	6,740	13 (10- 18)
All Cancers (females)	8,700	1,740	51 (38- 70)

* Cancer Doubling Doses from Kneale, Stewart, Mancuso, "Re-analysis of Data Relating to the Hanford Study of the Cancer Risks of Radiation Workers," Late Biological Effects of Ionizing Radiation, Vol. 1, International Atomic Energy Agency, Vienna, 1978.

Age-adjusted Doubling Dose for 0-15 year Wisconsin age group derived from 'Variation in Cancer Induction Per Rad with Age' (Table I in John W. Gofman, "The Cancer and Leukemia Consequences of Medical X-Rays," Osteopathic Annals, November 1975.) Ratio: Radiation Workers—1.5/Wisconsin Growing Child -- 7.2 (Annual dose distribution from age 0 to 15). Ratio rounded to: 1/5.

** ICRP classifications I and II: Bone marrow, thyroid, Lymph nodes, Reticular tissue, Pharynx, Lung, Pancreas, Stomach, and Large intestine.

Parentheses indicate range of Dose and Percentage Increased Risk.

risk from the estimated Annual Total Fission Dose accumulated (a state average increased risk of 10.3%).

Only 35 to 40% of the whole body radiation dose commitment to date is represented in the Annual dose accumulations on Table XXXV and discussion above. The risks will continue to increase each year as annual doses are delivered from radiation stored in bone, bone marrow and other cells of Wisconsin children. This will happen even if no new radiation is added. It is estimated that by the year 2000 (when the Growing Child group is 38 years old) the risks outlined above in Table XXXV will have increased by 40%. (Estimated on basis of age-specific factors for Annual doses to year 2000 from Sr-90 dose comcitments initiated in the 1963-1976 period.)

The increased risk assessment in Table XXXV does not include some important additional risks which were not quantified for lack of Wisconsin data. For instance, the Environmental Protection Agency, in February 1974, published a study of Environmental Radiation Dose Commitment: An Application to the Nuclear Power Industry in which a prediction was made of 30,000 health effects by the year 2020. The prediction was made on the basis of the release of only 4 radionuclides: Iodine-129, Tritium, Krypton-85 and the Actinides (plutonium, etc.) under normal operating conditions from atomic power plants. These radionuclides were chosen by EPA because of their very long half-life and their wide dispersion over large population groups. Of the 4 radionuclides in this study, only one, tritium, has been considered in the Wisconsin case-study. Tritium (H^3) is not included in the R3 Diet Dose since it is not monitored in Wisconsin, but in the Total Fission Dose it constitutes less than 2% of the total dose in accordance with ratios in UNSCEAR studies. (80)

The importance of other unmonitored radionuclides, diet, occupation and life-style, as well as testing infrequency and possible understatement of concentrations in monitored radionuclides have been discussed previously. All of these or combinations of these factors could result in much higher increased risk factors for groups and individuals within the Wisconsin populations.

Notes

1. It is unclear why the LaCrosse "Reference Station" was dropped from the Annual Environmental Radiation Survey reports for the Genoa reactor site at the end of June 1971 when milk monitoring reports through June 1972 continue for LaCrosse in the computer printouts for non-nuclear milk monitoring stations. The state Radiation Protection Section reported to the League Against Nuclear Dangers (LAND) in March 1975 that the LaCrosse station was dropped from computer printouts because "the dairy went out of business," yet, the the Annual Reports for the LaCrosse reactor report that the LaCrosse sample is from "the three major dairies in LaCrosse." (Emphasis added.) Monthly averages for radionuclide concentrations in milk at the "LaCrosse Reference Station"(in Annual nuclear site reports) and "LaCrosse" (in non-nuclear computer printout reports) vary in agreement. Some particularly large discrepancies occur: i. e., in May 1970, I-131 concentrations are reported as 19.26 pCi/1, 8 pCi/1,and 3 pCi/l for "LaCrosse," "LaCrosse Reference Station," and "Genoa" respectively. See Another Mother Fund for Peace (Beverly Hills, California) RADIATION MONITORING STUDY for more detailed information.

2. The only pre-1963 milk monitoring data available for Wisconsin from the state Radiation Protection Section is a report by the State of Minnesota for August through December 1962 which records concentrations of I-131 in milk ranging from "less than" 10 pCi/l up to 221 pCi/l. These results are many factors higher than reported in the 1963-1976 period under investigation in this study and strongly suggest that radiation doses in 1962 (a year of very active bomb-testing at the Nevada site) were many times higher than any reported in this study.

3. U.S. NRC Regulatory Guide 1.109, Table E-5, indicates a guide of 310 liters per year for Adults, 400 liters per year for Teenagers, 330 liters per year for Child, and 330 liters per year for Infant. NRC Regulatory Guide 1.42 (withdrawn) indicated 365 liters per year for Child. For consistency, 365 liters per year is used in this study for Infant-Child-Teen dose calculations; 330 liters per year for Adults. Since the methodology does not account for milk consumption through Dairy Desserts, butter, cream or dry milk products, the consumption factors assumed for young people in a major dairy state are considered a conservative estimate.

4. Reports of milk monitoring for nuclear stations in Annual Environmental Survey reports are given for monthly averages only. Generally, pooled milk samples are tested monthly: individual farm samples may be tested only quarterly. NRC Regulatory Guide 1.42 (withdrawn) would have required weekly testing for I-131 at individual farms near reactors.

5. The influence of formalin in reducing I-131 test readings has apparently been known at least since 1960: "The addition of formalin to milk causes elemental iodine to combine with protein and thus interferes with removal of iodine from the milk by ion exchange resins." (R. Scott Russell, Radioactivity and the Human Diet (Oxford, New York: Pergamon Press, 1966), p. 495). A comparison of I-131 milk monitoring by utilities at Point Beach – Kewaunee plants utilizing the chemical separation technique, to monitoring by the state laboratory, utilizing a gamma-scan technique, for a four-year-four-month period shows utilities reporting only half the contamination reported by the state laboratory. This discrepancy may have been due to the use of formalin in chemical separation techniques by the utility laboratory (at that time, Bio-Test Labs in Chicago). See Gertrude A. Dixon, "Statement to the Wisconsin Radiation Protection Council for the League Against Nuclear Dangers," (LAND: Stevens Point, Wis., Jan. 3, 1975).

6. Computer printout reports of milk radioactivity with exact weekly or bi-weekly concentrations were not received by the LAND/LEAF Research Team until October 1976.

7. All of these stated detection limits apply to gamma-scan analyses. Variations also occur for the chemical extraction testing process in Annual Environmental Survey Reports by the state laboratory for nuclear sites. Results of chemical extraction testing are not used in this study due to probable underestimation of concentrations of I-131. (See Note 5 above)

8. Letter from Chief, Radiation Protection Section to Research Director of LAND (March 12, 1975): "milk data . . . noted as less than 10 pCi/l could be noted as "0" or not detectable."

9. U.S. Environmental Protection Agency, EPA Assessment of Fallout in the United States from Atmospheric Nuclear Testing on September 26 and November 17, 1976 by the People's Republic of China, August 1977, p. 62. See also comments in same EPA study that I-131 samples should be analyzed "as soon as they arrive" to prevent loss of I-131 (8-day half-life), and discussion concerning long lag periods by Wisconsin laboratory between sampling and analysis in Another Mother

Fund for Peace RADIATION MONITORING STUDY.

10. See development of guidelines for "design objectives" in U.S. Nuclear Regulatory Commisssion, WASH-1258.

11. Thyroid doses are derived by R3 Diet Dose Methodology as outlined in Table I.

12. In addition to the dose to thyroid from foods containing I-131, doses from inhalation and external dose must also be considered particularly for populations near nuclear facilities. Also many other radionuclides released by nuclear fission produce a direct thyroid dose — in Hoenes and Soldat, Age-Specific Radiation Dose Commitment Factors for a One-year Chronic Intake (NUREG-0172, November 1977), thirty radionuclides are listed with specific numerical guides for impact on the thyroid. Although I-131 may be the largest single pathway for thyroid dose, it is by no means the only pathway.

13. For a discussion of the EPA Cross-check programs relating to monitoring in Wisconsin, see the Another Mother Fund for Peace RADIATION MONITORING STUDY.

14. U.S. EPA-520/5-88-002. See Note 9 for full title.

15. Many negative values appear also for I-131 but not as frequently as for Cs-137.

16. Computer Printout, Wisconsin State Laboratory, Milk Data, provided by George B. Morgan, Director U.S. EPA Environmental Monitoring and Support Laboratory, Las Vegas, Nevada, to J. D. Littlejohn, Research Assistant, Another Mother Fund for Peace, Stevens Point, Wisconsin, January 11, 1978. See analysis in Another Mother Fund for Peace RADIATION MONITORING STUDY.

17. The lack of Sr-90 testing inthe summer months of 1972 is difficult to understand since I-131 and Cs-137 levels were equal to or above the trend. Numerous weekly readings for I-131 above 10 pCi/l appear during the period of no Sr-90 testing which contradicts the advise of EPA that, if I-131 tests are greater than 10 pCi/l, Sr-89 and Sr-90 tests should be performed. See EPA reference cited in Notes 9 and 14 above, p. 11.

18. EPA-680/4-75-007, June 1975, p. 7.

19. Garner, op. cit., p. 22.

20. Eisenbud, op. cit., p. 37.

21. Garner, op. cit., pp. 37-38.

22. Garner, op. cit., p. 15.

23. Note that these current pathway figures depict only cow's milk being transmitted to humans — they do not indicate meat. See: NRC, Final Environmental Statement Related to Construction of Tyrone Energy Park (April 1977), Fig. 5-4: Exposure Pathways to Man. Evidence of a general disregard for radiation exposure from meats in the vicinity of nuclear generators is exemplified by a statement by the Chief of the Radiation Protection Section (Wisconsin): The Research Director of the League Against Nuclear Dangers (LAND) asked the Chief if he were considering monitoring meats in the vicinity of nuclear reactors as recommended by the Environmental Protection Agency, Environmental Radioactivity Surveillance Guide (ORP/SID 72-2, June 1972) — the response was: "What guide? . . . There wouldn't be any meat." (Telephone communication, October 22, 1974).

24. Garner, op. cit., p. 22.

25. Eisenbud, op. cit., p. 130. Note that this factor also applies to transfer of radionuclides to infants from human milk.

26. U.S. Atomic Energy Commission, Fallout from Nuclear Tests, Revised Ed., June 1966.

27. It is common practice for rural residents to utilize natural barnyard fertilizer to enrich the same garden plot for vegetables or fruits year after year. Accumulations of long-lived radionuclides could result in potentially extremely radioactive-rich garden soil and concommitant radiation exposure from consumption of home-grown foods. Doses could be much higher than indicated by deriving expected Total Diet dose from milk monitoring data alone which is the procedure in this methodology. Field studies are needed to explore this dose pathway before any dose quantification is possible.

28. Garner, op. cit., p. 22. Garner notes, further, that the researcher concluded that "'milking techniques used in experiments were at least as good as those used in commercial dairy herds' so that the concentrations found can be taken as an indication of the minimum concentration likely to occur."

29. Note that recent research has also revealed that chlorination of drinking water may enhance the hazards from plutonium in

the water: R. P. Larson, R.D. Oldham, Argonne National Laboratory, "Plutonium in Drinking Water: Effects of Chlorination on its Maximum Permissible Concentration," Science, September 15, 1978.

30. Russell, op. cit., pp. 402-403.

31. Ibid.

32. Garner, op. cit., p. 20. Garner points out that the lack of a root system, a 10-fold greater surface area than for grasses, and a growth period of tens of years are factors promoting accumulation of radionuclides in lichens.

33. U.S. Bureau of Radiological Health, "Dresden Study Data," as reported in Honiker vs. Hendrie: A Lawsuit to End Atomic Power (Summertown, Tennessee: The Book Publishing Company, 1978), p. 65. Concern that snow may be a more effective concentrator of radioactivity in fallout than rain was reviewed in: "Snow Pollution is Health Danger," Wisconsin State Journal, December 29, 1974. See also observation by D.E. Barber that higher milk radioactivity levels in Northern Minnesota than Southern Minnesota may be due to heavier snowfall in D. E. Barber, "Natural Differences in Concentrations of Strontium 90 in Milk," in Environmental Surveillance in the Vicinity of Nuclear Facilities — Proceedings of a Symposium Sponsored by the Health Physics Society, Edited by William C. Reinig. (Charles C. Thomas, 1970), pp. 368-381 passim.

34. "State Deer Harvest Could Be Biggest Ever," The Pointer, Stevens Point, Wisconsin, November 16, 1978.

35. Eisenbud, op. cit., p. 135.

36. Eisenbud, op. cit., p. 296.

37. U.S. Department of Health, Education and Welfare, "Preliminary Radiological Surveillance Studies at a Boiling Water Nuclear Power Reactor," December 1969, p. 163 in Hearings Before the Joint Committee on Atomic Energy: Environmental Effects of Producing Electric Power, October-November, 1969, pp. 824 ff.

38. Ibid., p. 168.

39. Eisenbud, op. cit., p. 487.

40. Garner, op. cit., p. 38.

41. Russell, op. cit., p. 339.

42. Garner, op. cit., p. 40.

43. Garner, op. cit., p. 22, p. 29.

44. Garner, op. cit., p. 30.

45. U.S. Department of Health, Education and Welfare, Radiological Health Data, November 1964, p. 549. See also: Kulp and Schulert, op. cit., p. 145, Table I-29 (Calcium and Sr-90 in New York Diets: Average of samples taken 3/60, 6/60, 10/60, 2/61, 5/61) listing 19 calcium-contributing foods -- only 5 categories contain less picocuries Sr-90 per gram Calcium than milk. These are shell fish, meat, fresh fish, poultry and eggs. Since this sampling is based on an urban population and since type of meat, particularly, is not specified, milk may rank second or third from bottom in Sr-90 concentrations for the Wisconsin rural population. (See preceeding discussion.)

Other foods not listed in these studies are potentially high radiation exposure constributors in Wisconsin diets. For instance, Mushrooms, according to Garner, are high concentrators and they are gathered wild by many ethnic groups in the state. Nuts are not produced in quantity for human consumption but they are an important animal food and will affect, particularly, wild game -- over a million squirrels were harvested in Wisconsin in 1975 and deer feed on wild nuts.

46. Garner, op. cit., p. 16.

47. Russell, op. cit., pp. 152-153.

48. Garner, op. cit., p. 16. See also Russell, op. cit., p. 101. The Wisconsin Department of Natural Resources has been concerned that chopping up old windbreaks in central Wisconsin to make room for large fields has promoted a condition in which "up to 10 tons of sandy soil per acre per year is being lost to the winds." ("On Wisconsin," The Milwaukee Journal, October 29, 1978). This practice will promote resuspension of radionuclides otherwise unavailable to humans. Extensive sand and gravel mining operations will also promote resuspension. The Milwaukee Journal article also observes that "developers and farmers often drain or fill up 'useless' swamps and bogs for building sites or cropland." Cesium and other radionuclides stored in these natural environmental "sinks" can re-enter the human food chain and produce internal or external doses when marshlands are disturbed.

49. Garner, op. cit., p. 239. D.E. Barber (op. cit., p. 376) points out that, in poor pasture, radionuclides can be "resuspended by hooves and wind."

50. Garner, op. cit., p. 33.

51. Garner, op. cit., p. 16.

52. "USDA Would Relax Boneless Meat Rules," Wisconsin Rapids (Wis.) Daily Tribune, October 6, 1977.

53. Garner, op. cit., pp. 37-38.

54. Eisenbud, op. cit., p. 128.

55. Garner, op. cit., p. 37.

56. Garner, op. cit., p. 44.

57. Eisenbud, op. cit., p. 128.

58. Garner, op. cit., p. 16.

59. U.S. Department of Health, Education and Welfare, Radiological Health Data, November 1964.

The relation of animal feed to radioactivity in cow's milk is explored in D.E. Barber, op. cit., pp. 368-381. Barber notes that wild grasses and "sedges in slough and marsh areas" are fed to dairy cows and cut as hay for winter feed. In a humid, marshy, organic soil radionuclides become more available to vegetation: "Marsh sites showed significantly more Strontium 90 than either upland or lowland sites . . . . Apparently, both moisture and organic content of the soil play an important role in explaining the higher concentrations of Strontium 90 in vegetation and soils from marsh sites. Almost without exception, vegetation on marsh sites had a higher Strontium 90 to calcium ratio. . . . Farms with high Strontium 90 concentrations in milk had a greater proportion of land in marsh than farms with low Strontium 90 concentrations in milk."

It is not known whether milk test samples in Wisconsin, used for dose estimate purposes in this study, are taken from upland dairy cows or marshland cows. Actual doses to individuals may be measurably affected by this unknown factor.

60. Russell, op. cit., pp. 330-331. Concentrations could be especially high where the same garden plot is fertilized, year after year, with radioactive-rich barnyard fertilizer.

61. For a review of radiation doses to Wisconsin nuclear power plant workers, see: Gertrude Dixon, "Comments on Advance Plans Submitted by Utilities to the Wisconsin Public Service Commission," delivered at Wisconsin Public Service Commission Hearings, Wisconsin Rapids, Wisconsin, April 1, 1977. For most recent information, consult: U.S. Nuclear Regulatory Commission, *Occupational Radiation Exposure at Light Water Cooled Power Reactors* (Annual Reports) and annual *Occupational Radiation Exposure* reports.

62. Garner, *op. cit.*, p. 19.

63. An underestimation of Cs-137 in milk will result in a proportional underestimation in the rest of the diet and external Cs-137 doses since, in this methodology, the total Cs-137 dose is based on the only available monitoring data which is milk. It is unfortunate that, for this radionuclide in particular, other environmental samples are not collected and tested.

64. Garner, *op. cit.*, p. 44.

65. John W. Gofman, "Radiation Doses and Effects in a Nuclear Power Economy: Myths vs. Realities," April 1976. CNR Report 1976-2.

66. Eisenbud, *op. cit.*, p. 379.

67. U.S. EPA, *Radiological Quality of the Environment in the United States, 1977*, Table 4-10.

68. See U.S. NRC, *Regulatory Guide 1.109* for age-specific dose calculations.

69. Correspondence from William L. Lappenbusch, Criteria and Standards Division, Office of Drinking Water, U.S. Environmental Protection Agency to Research Director, League Against Nuclear Dangers, Stevens Point, Wisconsin, March 27, 1978.

70. The degree of understatement of actual Annual dose by a methodology which considers ONLY ingestion of Cesium-137 in dairy products for one year (discounting total body dose from OTHER radionuclides, from OTHER pathways and from Annual dose delivered from commitments from OTHER past years) is illustrated in correspondence from Chief, Radiation Protection Section, Wisconsin Department of Health and Social Services (March 12, 1975) to Research Director, League Against Nuclear Dangers: "I conclude that the whole body dose [La-Crosse area — July 1969 to June 1971 period] is less than 0.14 millirems."

The actual Annual doses delivered to a Growing Child in the LaCrosse area in this period are: 1969 — 28.6 millirems; 1970 — 29.79 millirems; 1971 — 30.72 millirems. These doses are from ingestion only of Sr-90, Cs-137 and I-131 (the R3 Diet Dose). See Table XXIII. The factor of understatement of dose is over 200.

71. See Table I in U.S. Atomic Energy Commission, Fallout from Nuclear Tests, June 1966.

72. R. Scott Russell, op. cit., p. 197.

73. Garner, op. cit., p. 37.

74. Garner, op. cit., pp. 37, 40.

75. The Elk River (Minn.) reactor which was decommissioned in 1967 is not included in Table XXXI.

76. U.S. EPA, Assessment of Fallout in the United States from Atmospheric Nuclear Testing on September 26 and November 17, 1976 by the People's Republic of China (August 1977), p. 1.

77. U.S. EPA, Radiological Quality of the Environment in the United States, 1977, Table 5-16.

78. See D.E. Barber, op. cit., pp. 368-381.

79. See Kneale, Stewart, Mancuso, Reanalysis of Data Relating to the Hanford Study of the Cancer Risks of Radiation Workers, presented at the International Atomic Energy Meeting, Vienna, Austria, March 13-17, 1978.

80. See Table 4-9 in U.S. EPA, Radiological Quality of the Environment in the United States, 1977.

Section C:

Summary and Recommendations

Summary

Although there is general scientific agreement that some increased health risks and genetic damage must be expected from any added radiation dose, an "allowed" level has been permitted to the public by federal agencies from the production of nuclear weapons and nuclear-generated electricity. Adverse health effects are expected at these legally allowed dose levels. In 1972 the National Academy of Sciences (NAS) urged that legal exposure limits be reduced to a level one-third of the current limit to the public and one-tenth of the annual accumulated limit to nuclear workers. The NAS expected increases in the incidence of cancers, genetic damage and "ill health" even at this recommended lower level (5 rems in 30 years from sources of radiation except natural background and medical X-rays). (1) Dr. Karl Z. Morgan, eminent health physicist, has urged that X-ray exposure should be included in the legal limit. (2) More recently, the U.S. Environmental Protection Agency has proposed a ceiling of 25 millirems per year to the public from nuclear power – a 20-fold reduction of the legal limit. (3) The Natural Resources Defense Council and others have urged a 10-fold reduction of the presently allowed 5,000 millirems per year to nuclear workers. (4)

In the meantime, a body of scientific evidence is mounting demonstrating that doses at these "allowed" levels are causing more cancers than expected. The evidence relates to exposures of soldiers and civilians to bomb-testing fallout, nuclear workers in federal weapons facilities and nuclear submarine shipyards, children of parents exposed to X-rays, civilians in the vicinity of nuclear reactors. (5)

In view of the mounting evidence of increased risks at low levels of radiation, it becomes important that the ACTUAL dose to REAL individuals be determined as precisely as possible.

Methodologies for estimation of radiation dose by government agencies are based on expected releases from facilities. Computerized models predict environmental dispersion of radiation, particular pathways of exposure and final doses to humans. The accuracy of such models is doubtful. Recently a report by a group of fourteen Heidelberg (Germany) scientists who were independently asked to examine this methodology reveals that exposures to those living in the vicinity of nuclear reactors had been underestimated anywhere from 10 to 10,000 times. (6) Prediction of ultimate human radiation dose from records of amounts released assumes a high degree of accuracy concerning measurement of all radionuclides released, the pathways of dispersion and the bio-accumulation and uptake of each radionuclide. Frequent references in the scientific literature to limits of technological capability to measure releases and gaps in knowledge of the behavior in the environment of radionuclides contradicts the confidence placed on these predictive models. (7)

Actual field measurements of fission products in the food chain is sparse in recent scientific literature, except for milk monitoring. Calculation of radiation dose from concentrations in milk in the vicinity of nuclear reactors has revealed doses much higher than predicted by models based on expected releases. (8) Although the Nuclear Regulatory Commission requires some monitoring in the environment, this monitoring is not used to calculate actual doses:

> Environmental monitoring is not used as the basis for determining the potential dose to individuals but rather as a backup to the effluent monitoring program, as a means of public assurance, and as an indicator of the general applicability of the models . . . it would

be impossible for routine monitoring programs to determine actual doses to real individuals. (9)

Yet these same monitoring programs are included in all environmental impact statements for nuclear facilities as assuring proof that the low doses, predicted by models, to people in the vicinity will not be exceeded.

Most of the releases from these particular facilities are expected (by model prediction) to be restricted to a prescribed "impact" zone (usually a 50-mile radius). Within this area, the resulting dose is expected to be equally distributed to the whole area population. On the other hand, doses from weapons testing fallout are expected to be distributed equally to worldwide populations with differentiations reported only for broad latitudes. These "population" dose estimates are contradicted by long-established evidence that radiation is not distributed uniformly -- that nationwide doses from weapons testing fallout differ by factors of more than ten and that "first pass" citizens near reactors will receive higher doses. (10)

In view of these logical discrepancies inherent to current dose estimate methodology, the Land Educational Associates Foundation (LAND/LEAF) research team decided to develop and test a fission dose methodology based on actual environmental monitoring records in a particular geographical area. Such a realistic estimate of radiation exposure to actual residents of a specific area could serve as a better assessment of radiation-associated risks than the amorphous model-derived "general population" estimates.

The research team found that milk radioactivity was monitored in Wisconsin by the Radiation Protection Section. No other foods were monitored by the state and only three of the major fission radionuclides were measured even in milk. However, it was possible to estimate the total diet content of these three contaminants with established data on proportions of dairy products to the total diet. The basic calculation, derived from actual local monitoring, was called the R3 Diet Dose (3-Radionuclide Diet Dose). Additional proportions for external dose were available in United Nations and EPA studies for estimation of the Total Fission Dose (See SECTION A: METHODOLOGY.)

In Wisconsin milk monitoring records were available for a 14-year period for 6 monitoring stations. Since NRC guides are available for deriving the millirem dose for each of 189 radionuclides for specific age groups, it was possible to calculate the lifetime accumulated dose for a Wisconsin teenager born at the start of the monitoring period (1963). A calculation was also made for the dose during those 14 years to an Adult. No state data was available for estimation of dose to those born in the early 1950's who would have accumulated the highest dose, having grown up during the atomic era. The research team estimated that this group (18 to 25-year-olds) would have accumulated a fission dose approximately 50% higher than those born in 1963.

In order to stay with data which could be absolutely verified by state monitoring records, the research team had to omit many radiation sources which had not been monitored and could not be accurately estimated. The dose findings are, therefore, not the WHOLE dose. The conclusions are understatements of the actual dose particularly because of the following necessary omissions: (11)

1. The R3 Diet Dose includes ONLY the 3 monitored radionuclides (Strontium-90, Cesium-137 and Iodine-131) out of a possible 189 listed by the NRC. Studies have definitely established the presence of many other radionuclides in foods. Twelve others have been identified in milk alone.

2. Only one food, milk, is monitored by the state despite the fact that many other foods are known to absorb higher concentrations of radioactivity. Milk has been monitored because of the relative ease of collecting and testing, because it reflects changes in radiation levels more quickly and because large numbers of people may be affected by contamination. In a list of 19 calcium-contributing foods in the U.S., cow's milk ranks 13th for Sr-90 concentration. Scientists observed 15 years ago that people who derive most of their calcium from foods other than milk will have higher doses from Sr-90 than those who "dilute" their calcium diets with milk. The

Atomic Energy Commission described the cow as a "decontaminator" in the radiation pathway to humans since only a tenth of the radioactivity in the grass it consumes is secreted in its milk. Cs-137 concentrations are higher in meats, particularly venison, many fruits and vegetables and 25 times higher in goat's milk than in cow's milk. Foods especially high in radioactivity include potatoes, whole wheat, bananas, leafy vegetables, soybeans, berries, venison, nuts, cheese, cabbage. None of these foods are monitored by the state laboratory. Fresh vegetables will yield higher doses from short-lived nuclides than calculations assuming "store life" for dose from vegetables. Researchers in the early 1960's had already concluded that vegetarians and people who consumed primarily locally produced foods would have much higher radiation doses than the "average" urban population. None of these additional dose considerations were quantifiable in the LAND/LEAF dose estimates because of lack of monitoring data.

3. Drinking water is not included. It could be a major source of radiation for some individuals. Farm runoff, for instance, is recognized in the state as a major source of drinking water pollution. Concentrations of radioactivity in soil runoff are ten times higher than in residual soil. And, since the manure and urine of cows contains 10 times as much radioactivity as the milk, barnyard runoff into drinking water reservoirs or radiation-rick "natural" fertilizer applied year after year to family garden plots could expose humans to much higher doses.

4. Failure of the state to monitor milk with regularity, particularly Sr-90 in the summer months, could minimize the final estimates. The technical ability of the laboratory to detect the levels of contamination in milk is questionable — for instance, an EPA cross-check indicates that the laboratory has a tendency to underestimate the Cs-137 in milk except at very high levels. Cross-check samples were underestimated by up to 90%. The laboratory could obviously have sampled foods expected to have higher concentrations (such as goat milk or venison) with better precision. More than 150,000 deer were harvested in Wisconsin last year but no meats are moni-

tored by the state.

5. Because of a lack of consumption and radioactive concentration data, the research team omitted all dairy products except milk and cheddar and cottage cheese. Dairy desserts, particularly, are consumed in large quantities in this state and would contribute additional doses.

These major limitations and omissions insure a large factor of conservatism in the LAND/LEAF dose estimates.

First, the research team calculated a "50-year Dose Commitment" using NRC age-specific dose guides. "Commitment" means exactly that: an iron-clad contract which spells out just how long each radionuclide will take up residence in the body and how many millirems of radiation dose it will deliver before it decays into a new generation of "daughter" radionuclides and finally dies. Humans can create these previously non-existing radionuclides but they CANNOT destroy them nor can they shorten their pre-ordained lifespan. The body-cell lifetime of radionuclides varies from a few seconds to over 50 years — many outlive the body itself. Some do their damage quickly and die or are excreted — others take up residence in bone and tissue delivering only part of the dose "commitment" yearly, bombarding cells every day for years until their decay time, pre-ordained by the laws of physics, is completed. Some of the strontium-90 that found its way into a baby's milk bottle in the early 1950's is still in the cells of that individual. Every year it delivers an additional "annual" dose. Every year it increases that individual's risk of cancer.

When the LAND/LEAF research team had calculated the dose commitments for Wisconsin, a very wide range of difference for age and area of residence was apparent immediately. Residents in the north-west-central section had accumulated a dose nearly twice as high as residents in the south and east section due, primarily, to a marked difference in bomb-testing fallout before the test-ban treaty in 1963. And the Growing Child had accumulated a total body dose twice as high as Adults and a Bone dose three times as high as Adults. Obviously the "average population" dose numbers are meaningless even for residents of an area as small as Wisconsin.

The average Total Body Dose Commitment for only three radionuclides in food (the R3 Diet Dose) ranged from 25.41 millirems per year to an adult in South-southeastern Wisconsin to 100.86 millirems per year to a growing child in Northwestern Wisconsin. These doses are from 4 to 16 times higher than United Nations estimates for the Northern Latitude from bomb-testing fallout. The estimated average Total Fission Dose Commitment to Total Body ranged from 59.4 millirems per year to above adult up to 228.3 millirems per year to the above growing child – from 5 to 14 times higher than expected for the Northern Latitude.

The average Dose Commitment to Bone from the R3 Diet only ranged from 97.7 millirems per year up to 404.65 millirems per year. These estimates are 12.3 to 51 times higher than United Nations estimates. The estimated average Total Fission Dose Commitment to Bone ranged from 192.9 millirems per year up to 791.8 millirems per year – higher by a factor of 14.4 to 42.6 than United Nations estimates of bone dose in the Northern Latitude.

The LAND/LEAF research team also calculated the part of the 50-year commitment dose which had already decayed each year and delivered its "annual" radiation effect. This average "annual" fission dose during the 14-year period was compared to natural radiation dose or X-rays which deliver an average annual dose of 100 to 125 millirems.

The Whole Body radiation dose to Wisconsin citizens from just 3 radionuclides in food (the R3 Diet Dose) ranged from 10% to 36% of the annual X-ray or natural radiation dose. The R3 Diet Dose to the Bone ranged from a third of the natural or X-ray dose for a southeast Wisconsin adult up to one and a half times more than the natural dose for a child in northwestern Wisconsin. These doses were from only 3 radionuclides in food as indicated by monitoring by the state Radiation Protection Section. The estimated Whole Body Total Fission Dose ranged from 22% to 81% of the natural dose per year. The annual average Total Fission Dose to Bone (an estimated 28% of the Commitment Dose) was much higher ranging up to twice the annual X-ray or natural radiation dose.

All of these estimates of annual radiation exposure from nuclear fission are much higher than model-derived estimates for the general population of not more than 4% of natural radiation dose.

The current increased risk of radiation-induced cancers to Wisconsin 14-year-olds was calculated on the basis of "annual" doses already delivered from decayed radioactivity by 1978. Risk estimates derived from the recent ERDA-sponsored study of cancer deaths among nuclear workers and an age-adjusted relative risk scale were used. In this analysis, a 100% increased risk is equal to double the normal cancer risk.

The radiation dose just from three radionuclides in their food has already increased the risk of myeloma and myeloid leukemia for Wisconsin 14-year-olds by 41% to 74% depending upon their area of residence. For the Total Fission Dose the increased risk was estimated to be from 93% to 168% higher than the normal risk.

For all cancers classified by the International Commission on Radiation Protection as High Sensitivity cancers for radiation-induction the current increased risks for this group ranged up to 20% from just the 3 radionuclides in food — up to 44% from the estimated Total Fission Dose. Cancers of the bone marrow, thyroid, lymph nodes, reticular tissue, pharynx, lung, pancreas, stomach and large intestine are classified as High Sensitivity cancers.

The average increased risk of ALL cancers was 13% (up to 18%) from the Total Fission Dose based on the risk estimate for males in the ERDA-sponsored study. On the basis of a smaller sample of females this study estimated that their cancer risk from radiation was much larger. Applying this less certain risk estimate to 14-year-old Wisconsin females, the LAND/LEAF research team calculated an average increased risk of all cancers of 51% (up to 70%) from the accumulated annual Total Fission Dose.

Dr. John W. Gofman's estimate of a doubling-dose of 43.5 rems for all cancers was also applied to accumulated annual doses for the Wisconsin Growing Child group. Adjusted for age at irradiation, the Gofman doubling-dose yields from 3.38% to 6.15% increased cancer risk from the accumulated Annual R3 Diet Dose (a state average increased risk of 4.54%); and, 7.7% to 13.94% increased risk from the estimated Annual Total Fission Dose accumulated (a state average increased risk of 10.3%).

As alarming as these figures are, these young people are already COMMITTED to even higher risks, since only 35% to 40% of the radiation absorbed by their body cells during their lifetime (the "dose commitment") had decayed and delivered its risk effect by 1978. The rest will CONTINUE to radiate, delivering MORE dose constantly for the rest of their lives. Their risks of cancer will continue to rise with every disintegration of the restless radionuclides imprisoned in their body cells. This will happen EVEN IF NO NEW RADIATION IS ADDED. The future radiation doses each year can be numerically predicted NOW with available NRC dose guides. With fresh radiation doses, externally or internally, their risks will rise more steeply and for a longer time.

In an effort to identify sources of the Wisconsin radiation exposure the LAND/LEAF research team investigated the most evident possibilities of bomb-testing fallout and releases from midwestern nuclear facilities. Definitive evidence of exact sources is, of course, impossible in the absence of exact amounts released, dates, meteorological conditions, environmental monitoring. However, the major sources can be pointed out over a period of years.

In the 14-year period investigated in Wisconsin the milk monitoring clearly shows bomb-testing fallout residual all over the state in 1963 with heavier concentrations in the north-west-central portion. Chinese weapons testing added to the dose in 1965 and 1966. In the spring of 1970 the monitoring shows a sharp rise all over the state not directly attributable to any bomb testing. No one investigated. The Dresden (Illinois) reactor to the south was releasing more radiation every year, reaching over a million curies by 1970. The Genoa reactor was just beginning to operate. An "underground" bomb test in Nevada inadvertently leaked. These are all "possible" sources.

By the late 1960's and early 1970's the monitoring shows Sr-90 levels greatly diminished. From early contamination well above national averages, concentrations declined to the national average. Differences between northern and southern Sr-90 levels narrowed indicating that high soil deposits from bomb-testing COULD BE OVERCOME WITH TIME and that new fallout was small. In fact, in 1971 and 1972 the state monitoring shows all stations below the national average for Sr-90. The reliability of the figures is, however, questionable since no tests were taken in many months — in 1972 no Sr-90 data was collected at most stations from April through September.

This hopeful declining trend in radioactive contamination of the state changed abruptly in the later years. In the 1970 to 1976 period 12 nuclear reactors in addition to the two existing ones began operating in Wisconsin or adjoining states. Suddenly, in 1973, Sr-90 levels shot up above the national average at all six stations and stayed above the average through 1976 (the latest data available and the close of the investigated period). Contamination levels were back up to concentrations which hadn't occurred for five or six years. Meanwhile, the national average continued to decline slowly. The state average for the four-year period is more than twice the national average for Sr-90 — 263% higher in 1974.

Nuclear waste was raining and snowing down on Wisconsin again.

Where did it come from? A check of all high Iodine-131 levels, indicating fresh fallout from recent releases, did not bear out a correlation with immediate fallout from Chinese bomb tests during this period to any significant extent. The highest levels of I-131, those above 10 picocuries per liter, do not correlate with bomb tests.

The excess Sr-90 in 1973-1976 was not evenly distributed. The Eau Claire and Green Bay stations showed the largest increase over the base year of 1971. In 1974 the Monticello (Minn.) reactor west of Eau Claire had the highest gaseous releases from reactors in the nation: in 1974 the Point Beach plant, near Green Bay, exceeded expected gaseous releases by a factor of two according to the EPA. On the other hand, the station which showed most improvement in milk contamination relative to other stations in the second half of the 14-year period was the central Wisconsin station, geographically furthest removed from the encircling reactors. But, even at this station, contamination was above the national average — evidence that nuclear wastes do not respect the diagrammed "impact zone" boundaries delineated by the NRC for reactors.

The LAND/LEAF research team did not have the data to identify exact sources of this recent recontamination of the state but it is obvious that the excess radiation dose over the national average is not due to residual bomb-testing fallout. It is possible that some originated at distant, up-wind nuclear weapons facilities or was delayed fallout from Chinese testing. The inconsistent distribution of the Sr-90 suggests more localized sources. For instance contamination in the Genoa reactor area diminishes in 1974 while the other stations (particularly Eau Claire and Green Bay) show increased radiation levels. On the other hand, in 1975 Sr-90 levels increased in the Genoa area while all other stations showed less contamination.

Some of the discrepancy between the expected U.S. average population dose and the 14-year average in Wisconsin is, no doubt, due to fallout before the test-ban treaty from Nevada bomb tests. Monitoring records in the early 1960's clearly show differential fallout with highest levels in the Northwest and lowest in the Southeast. However, even southern Wisconsin exposures are many factors above the expected average and large areas of the country received doses equal to or exceeding Wisconsin's from the Nevada tests. Figures XII and XIII show nationwide fallout patterns from just two Nevada tests — each test will have a different fallout pattern, depending upon wind direction, velocity and precipitation.

It must be concluded, then, that the understatement of actual dose to real individuals is not solely a Wisconsin phenomenon. Some factors which account for the discrepancy became obvious during the dose estimate research process.

A major factor in extant radiation exposure estimates which will always understate the dose to a child is the practice of using population-weighted age distributions. The small percentage of total population under age 10 (about 19% in Wisconsin) receives a much higher dose

FIGURE XII

STRONTIUM-90 FALLOUT FROM NEVADA TESTS

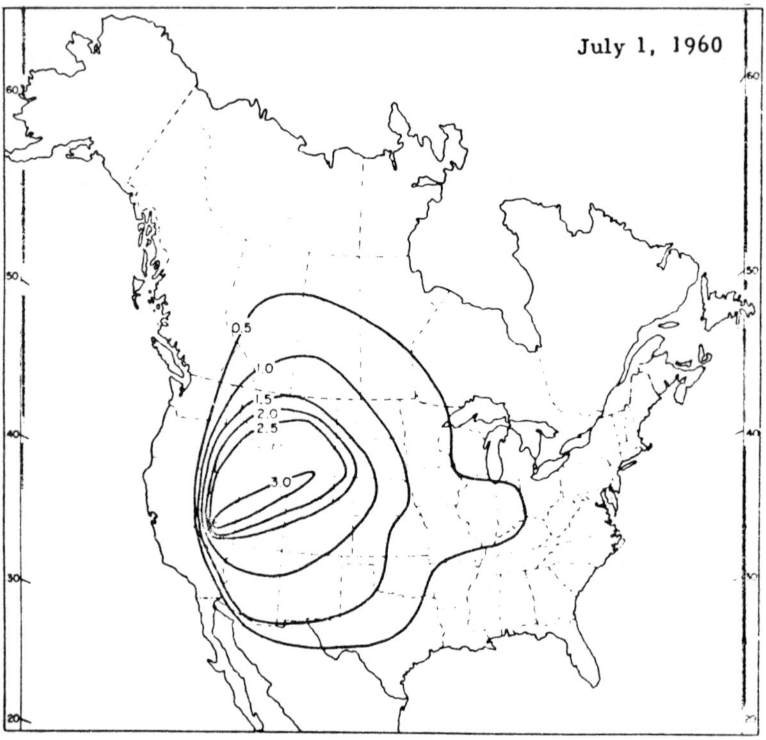

Strontium-90 deposition in U.S. and Cañada from Nevada Tests, July 1, 1958 and July 1, 1960. Contours show excess mc $Sr^{90}/mi^2/in.$ over latitude average.

FIGURE XIII

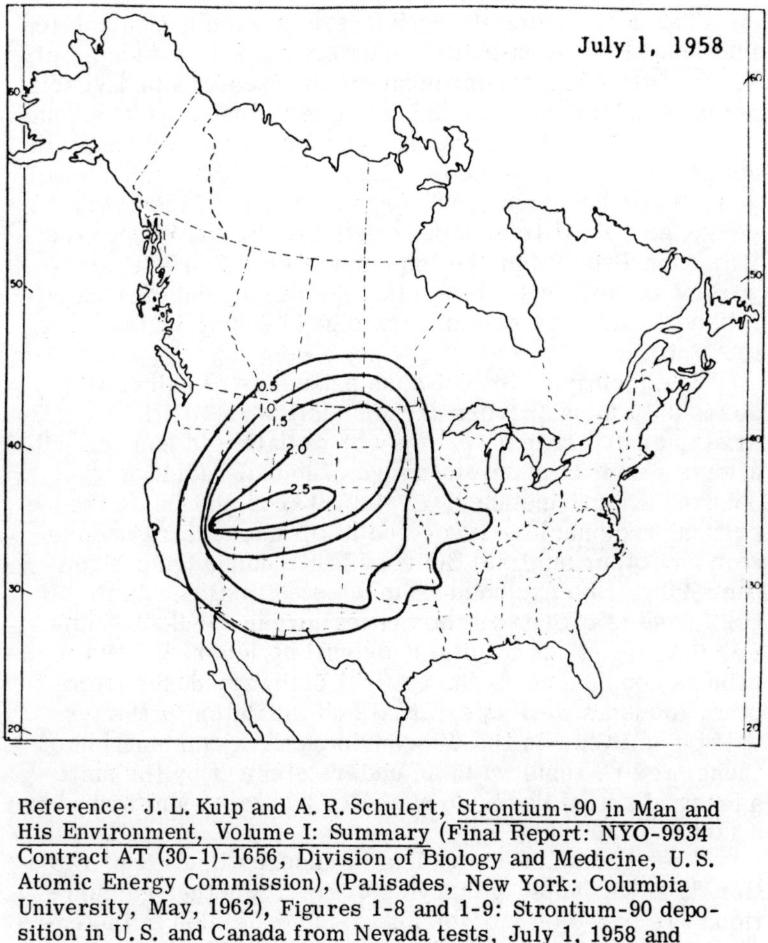

Reference: J. L. Kulp and A. R. Schulert, Strontium-90 in Man and His Environment, Volume I: Summary (Final Report: NYO-9934 Contract AT (30-1)-1656, Division of Biology and Medicine, U. S. Atomic Energy Commission), (Palisades, New York: Columbia University, May, 1962), Figures 1-8 and 1-9: Strontium-90 deposition in U. S. and Canada from Nevada tests, July 1, 1958 and July 1, 1960. Contours show excess mc Sr-90/mi^2/in. over latitude average.

than adults. These large infant and child doses are always diluted in any "average population" configuration. (In the Wisconsin case even adult doses are many factors higher than extant estimates.)

The dose "commitment" factor is often not calculated into annual dose estimates. For Sr-90, for instance, only 4.4% of the dose commitment due to exposure in a year is actually delivered as an annual dose in that year – annual doses will continue for years to come as the radionuclide continues to decay in body cells. These continuing annual doses must be added to any fresh exposure dose every year. An annual fission dose estimate by the Wisconsin Radiation Protection Section completely ignored this important factor, even though the monitoring data for calculation of past dose commitments had been collected by the state.

The factor in current dose estimates of calculating doses only for a "critical" radionuclide, a "critical" organ, or a "critical" pathway of radiation to humans will always result in understatement of actual radiation exposure. If, for instance, Cesium-137 is chosen as the critical radionuclide (expected to produce a larger dose than any other nuclide) the very large doses from Strontium-90 will be excluded from dose estimates. Again, if bone dose is chosen as the critical organ for Strontium-90, the large dose to total body will be ignored. And if milk is considered as the critical pathway, doses from other foods as well as external and inhalation pathways will be omitted. In the Wisconsin case, a combination of these factors resulted in an understatement by the state agency of actual dose (documented by state monitoring) by a factor of 200.

The above factors leading to underestimation of radiation dose will apply in varying degrees whether calculations are based on monitoring of radionuclides in the environment or on projected releases from facilities.

All of these considerations apply to estimates in the "scientific" literature as well as in estimates usually released to the public, primarily by licensees of nuclear generators. But the public understanding of actual radiation exposure is further minimized by the implication in public statements that the dose from some particular facility or some particular fission incident is the whole fission dose exposure. The sources of radiation exposure are many and may be difficult to identify but once the radionuclides have been absorbed by human cells the health damage is cumulative regardless of origin. If radionuclides are identified in the food chain by monitoring they constitute a real dose and a real threat to the health of human beings. The cumulative risk cannot be diminished or divided simply by denial of responsibility by either atomic weapons producers or nuclear facility operators. As long as radiation exposure calculations are expressed in terms of source or release of radiation, each part of the fission dose can easily be underestimated and the whole fission dose risk remains unknown. Only by basing dose calculations on radionuclides in the environment will the whole dose become apparent.

The findings of this study demonstrate that current radiation dose methodologies will seriously understate the actual dose to real people. A more realistic base for dose estimates to a population is the monitoring record of fission products in the environment.

Recommendations

The following recommendations are based on testing of the R3 Diet Realistic Dose Methodology in the Wisconsin Case Study:

1. This radiation dose estimate methodology should be applied to other populations throughout the world in order to ascertain realistic fission dose estimates. The accumulated fission dose estimate should be based on the available records of contamination in the environment, not on estimates of releases. The concept of a "general population" dose estimate must be discarded as meaningless in the light of a very wide range of actual dose found in this Wisconsin investigation.

2. An accumulated exposure estimate must be routinely updated every year on the basis of the monitoring record. In the event of any unusually high radiation exposure event, such as the 1979 accident at the Three Mile Island nuclear generator in Pennsylvania or fallout from bomb-testing, the additional dose should be calculated

from careful monitoring of all major pathways and by individual body counting and urine sampling. The accumulated dose from all past exposure must be taken into account in assessing risks to the population.

3. Dose differentials due to age, diet patterns, work related exposure, etc., should be made explicit. Separate dose and risk evaluations should be calculated for high risk individuals: i.e., the fetus, young children, children with a history of allergies, etc., nuclear facility workers, the aged and infirm, and any population exposed to especially high fallout from bomb-testing or reactor accidents. (12)

4. Radiation monitoring must be refined and expanded. The best analytical techniques should be used with consistency at all monitoring stations. (13) Foods other than milk must be monitored, particularly those known to be high concentrators of radioactivity and those which may be suspected to contribute large portions of the total radiation exposure to particular populations because of dietary patterns. Sampling should include areas of different soil types, precipitation patterns and radionuclide runoff pathways which are expected to yield variations in radionuclide concentrations. (14) The concept of an "average" radiation exposure in large geographical regions must be discarded as unrealistic. The exposure pathways of animal wastes and resuspension of radionuclides must be investigated for potentially large radiation doses to humans. (15) The monetary investment in improving radiation monitoring to the point where it can be used to reliably predict actual doses to real individuals must be considered in the cost accounting of producing nuclear weapons and nuclear-generated electricity.

5. All radiation exposure data must be freely available to any individual or group. This data must be presented in understandable terms of the degree of increased risk of radiation-induced disease and damage to the gene pool so that intelligent public decisions can be made by those affected by the increased risks. All conflict of interest must be eradicated from federal, state or institutional agencies involved in radiation monitoring or responsible for calculation of doses and risks. (16) All government agency personnel responsible by law for ra-

diation protection of citizens must be held accountable for the correctness of any quantified dose estimates. The rights of citizens to be informed of actual risks must be upheld.

6. Epidemeological studies should be undertaken to assess the physical assault of doses already delivered. The Wisconsin Case Study affords the basic data for the study of health effects on large population groups exposed to differential accumulated doses of radiation. In addition to a study of the incidence of leukemias and cancers in these populations, other expected health effects of radiation such as accelerated aging, increases in diseases related to genetic changes and appearance of debilitating diseases at earlier ages should be investigated. (17) Methods for improving collection and recording of data concerning incidence of diseases, rather than mortality data, must be refined.

7. The nuclear fission dose to people must be recognized for what it is: a nuclear waste problem. As such, it is germane to any deliberations concerning storage or containment of nuclear wastes from any facility. The quantity of nuclear waste "produced per unit of electricity generated is a fixed quantity, set by the immutable laws of nuclear fission physics." (18) The fission dose commitment researched in the Wisconsin case study is nuclear waste irrevocably stored in the body cells of real individuals. It has been "lost on the way to the bank" — it is irretrievable — it will never be safety contained in some ultimate waste storage repository. (19) It will inevitably affect the health of individuals in the Wisconsin case study and others yet unborn.

Notes

1. See National Academy of Sciences Report of the Advisory Committee on the Biological Effects of Ionizing Radiation, The Effects on Populations of Exposure to Low Levels of Ionizing Radiation, November 1972.

2. See U.S. Department of Health, Education and Welfare, 4th Annual National Conference on Radiation Control: "Save A Rad," April 30 — May 4, 1972.

3. The U.S. Environmental Protection Agency proposal of a legal limit of 25 millirems total body exposure and 75 millirems thyroid exposure per year from the nuclear fuel cycle is to go into effect in December 1979.

4. For a recent frank discussion of the difficulty of reducing radiation exposure to workers see comments by Robert B. Minogue, Director of the Nuclear Regulatory Commission's Office of Standards Development in "Radiation Roulette? Exposure Limits: Studies Cast Doubt on Adequacy," *The Day* (New London, Conn.), June 30, 1978: "Minogue said when the present limits were approved, many experts still believed there was a level of radiation — a 'threshold' — below which exposure was not harmful. 'What we have found is, by God, there ain't no threshold,' he said. 'There are some die-hards who still believe in it, but it's a myth that there is a threshold.'" The present exposure limit to workers is 5,000 millirems per year: "If the limit is changed to a half-rem a year — 500 millirems — or even 2.5 rems as others suggest, it would be difficult to find enough trained workers to do the job because they would be 'burned out' (reach exposure limit) quickly."

5. See *Honicker vs. Hendrie: A Lawsuit to End Atomic Power* (Summertown, Tennessee: The Book Publishing Company, 1978) for a comprehensive review of research on the health effects of low-level radiation. See also *Proceedings of a Congressional Seminar on Low-Level Ionizing Radiation* (A Report Transmitted by the Subcommittee on Energy and The Environment of the Committee on Interior and Insular Affairs of the U.S. House of Representatives, Ninety-fourth Congress, Second Session, November 1976. (U.S. Government Printing Office, 1976).

6. From "Tutorium Umweltschutz an der Universität Heidelberg, Radioökologisches Gutachten zum Kernkraftwerk Wyhl," May 1978 as reviewed in Jeannine Honicker, Petitioner before the NRC in the Matter of Petition for Emergency and Remedial Action, Comments of the Petitioner's Staff on the NRC Staff Response to her Petition, January 1979.

7. For a detailed technical analysis of the many uncertainties of predicting radiation dose from models based on releases of radionuclides from reactor facilities see: U.S. Environmental Protection Agency, "Review of the Draft Environmental Statement Concerning Proposed Rulemaking Action: Numerical Guides . . ." in U.S. Atomic Energy Commission, *Final Environmental Statement Concerning Proposed Rule Making Action: Numerical Guides for Design Objectives and Limiting Conditions for Operation to Meet the Criterion "AS LOW AS PRAC-*

TICABLE" for Radioactive Material in Light-Water-Cooled Nuclear Power Reactor Effluents (WASH-1258) Volume 3, Comments on the Draft Environmental Statement and Staff Responses Thereto, July 1973, pp. 255-309.

8. See especially: E.J. Sternglass, "Cancer Mortality Changes Around Nuclear Facilities in Connecticut," February 10, 1978. Barry Commoner, "Strontium-90," Hospital Practice, November 28, 1977.

9. Statement of Roger J. Mattson, Director, Division of Siting, Health & Safeguards Standards, U.S. Nuclear Regulatory Commission, presented at the U.S. EPA Hearing on Proposed 40 CFR Part 190, March 8, 1976. Attachment A, pp. 52-53.

10. See U.S. Environmental Protection Agency, Office of Radiation Programs, Environmental Radiation Dose Commitment: An Application to the Nuclear Power Industry, February 1974.

11. See this study Section B: Wisconsin Case Study for details and documentation of the following summary.

12. See report of Dr. Irwin Bross on Tri-state Leukemia study in American Journal of Public Health, February 1979. Also in "A New Report on Radiation," Boston Globe, February 1979.

13. In Wisconsin, monitoring and reports of monitoring are not the the same for non-nuclear and nuclear stations inhibiting comparison of all state areas. (See discussion relating to difficulties of assessing dose in the LaCrosse area after reports for the non-nuclear station were dropped in June 1972). For a comprehensive discussion of monitoring difficulties at the Genoa, Wisconsin reactor see the Radiation Monitoring Study by Another Mother Fund for Peace (to be published in 1979).

14. The importance of sampling a variety of foods grown under different conditions was emphasized in research in the early 1960's. For instance, Dr. W. B. Rankin, Assistant Commissioner, U.S. Food and Drug Administration noted: "I would like to sound a word of warning against too great a limitation of the collection and examination of food samples. We need to examine samples and to continue to do so to determine the facts upon which to make sound public health decisions. . . . There is no substitute, in food and drug work, to examining samples to determine the actual conditions existing. We sound a note of caution against a trend that has become evident here today -- a desire to curtail materially the number of food samples examined for radioactivity. We do not have enough knowledge of this area to warrant drastic limitation of samples,"

in Food and Agriculture Organization of the United Nations, FAO Atomic Energy Series, No. 5, "Agricultural and Public Health Aspects of Radioactive Contamination in Normal and Emergency Situations," Papers presented at the seminar, Scheveningen, the Netherlands, 11-15 December 1961 (FAO, Rome: 1964), p. 231.

15. See previous discussion concerning changes in radiation pathways figures omitting the exposure of humans to radiation through deposits of animal wastes on soils. Even the original pathways figures by the International Commission on Radiation Protection did include this pathway which is no longer to be found in Nuclear Regulatory Commission pathways figures. Exposure could be sizeable from this pathway since the cow, for instance, secretes 10 times as much radiation in its feces and urine than in its milk, and this material is deliberately used to fertilize vegetable gardens and fields. See Figure 1 and Figure 2 in ICRP Publication No. 7, Sept. 13, 1965.

16. The state of Wisconsin has not published accumulated dose estimates on the basis of its own monitoring record and available federal dose calculation guides. The Chairman of the Radiation Protection Council, holding that position since the inception of the Council in 1963 to the present, is a former Vice President and lobbyist for Wisconsin Electric Power Company, operators of the Point Beach nuclear reactor and co-owner of proposed new plants in the state.

17. For a discussion of the need for epidemiological research in these areas see Dr. Rosalie Bertell, "Brief Critique of the BEIR Report," September 1976.

18. John W. Gofman, On the Way to the Bank or Why There Will NEVER Be a Solution to the Radioactive Waste Problem, November 1977. C.N.R. Report 1977-7.

19. Ibid.

Bibliography

Allen, Robert E. (Division of Operational Safety, U.S. Atomic Energy Commission), Summary Information on Accidental Releases of Radioactive Effluent to the Atmosphere from Underground Nuclear Detonations Designed for Containment, August 5, 1963 – June 30, 1971 (Appendix A, WASH-1183, UC-41). Washington, D.C.: U.S. Government Printing Office, February 1974.

Another Mother Fund for Peace, RADIATION MONITORING STUDY. Beverly Hills, CA, To Be Published 1979.

Bertell, Rosalie, "A Brief Critique of the BEIR Report," September 1976.

Bross, Irwin, "Tri-State Leukemia Study," American Journal of Public Health, February 1979.

Carter and Moghissi, "Three Decades of Nuclear Testing," Health Physics, July 1977.

Commoner, Barry, "Strontium-90," Hospital Practice, November 28, 1977.

Correspondence from Lawrence J. McDonnell, Chief, Radiation Protection Section, Wisconsin Department of Health and Social Services to Gertrude A. Dixon, Research Director, League Against Nuclear Dangers, Inc., March 12, 1975.

Correspondence from William L. Lappenbusch (Criteria and Standards Division, Office of Drinking Water, U.S. Environmental Protection Agency) to Gertrude A. Dixon, Research Director, League Against Nuclear Dangers, Inc., March 27, 1978.

Dixon, Gertrude A. "Comments on Advance Plans Submitted by Utilities to the Wisconsin Public Service Commission," delivered at Wisconsin Public Service Commission Hearings, Wisconsin Rapids, Wisconsin, April 1, 1977.

________________. "Comments on Draft Environmental Statement Related to Construction of Tyrone Energy Park, Unit 1," submitted to the U.S. Nuclear Regulatory Commission, August 14, 1976.

________________. "Statement to the Wisconsin Radiation Protection Council for the League Against Nuclear Dangers, Inc.," delivered at Council meeting, January 3, 1975.

Eisenbud, Merril. Environmental Radioactivity, 2nd Ed. New York: Academic Press, 1973.

Fowler, John M., ed. Fallout: A Study of Superbombs, Strontium 90, and Survival. New York: Basic Books, Inc., 1960.

Garner, John R. (U.S. Environmental Protection Agency, Radiation Office). Transfer of Radioactive Materials from the Terrestrial Environment to Animals and Man. Cleveland, Ohio: The Chemical Rubber Co., 1973.

Gofman, John W. "The Cancer and Leukemia Consequences of Medical X-Rays," Osteopathic Annals, November 1975.

______________. On the Way to the Bank or Why There Will Never Be A Solution to the Radioactive Waste Problem (C.N.R. Report 1977-7), November 1977.

______________. Radiation Doses and Effects in a Nuclear Power Economy: Myths vs Realities. (C.N.R. Report 1976-2), April 1976.

Hoenes, G.R., and J.K. Soldat. Age-Specific Radiation Dose Commitment Factors for a One-year Chronic Intake (NUREG-0172 – U.S. NRC Contract No. B21446). Richland, Washington: Battelle Pacific Northwest Laboratories, November 1977.

Honicker vs. Hendrie: A Lawsuit to End Atomic Power. Summertown, Tennessee: The Book Publishing Company, 1978.

Honicker, Jeannine (Petitioner before the NRC in the Matter of Petition for Emergency and Remedial Action). Comments of the Petitioner's Staff on the NRC Staff Reponse to her Petition. The Book Publishing Company, January 1979.

International Atomic Energy Agency, Biological and Environmental Effects of Low-Level Radiation, 3 volumes (Proceedings of a

Symposium, Chicago, 3-7 November 1975, Organized by the IAEA and WHO). Vienna: International Atomic Energy Agency, 1976.

International Commission on Radiation Protection, ICRP Publication No. 7, September 13, 1965.

Kneale, Stewart, Mancuso. Reanalysis of Data Relating to the Hanford Study of Cancer Risks of Radiation Workers. Presented at the International Atomic Energy Meeting, Vienna, Austria, March 13-17, 1978.

Kulp, J. Laurence and Arthur B. Schulert. Strontium-90 in Man and His Environment, Volume I: Summary (Final Report: NYO-9934 Contract AT (30-1)-1656, Division of Biology and Medicine, U.S. Atomic Energy Commission). Palisades, New York: Geochemical Laboratory, Lamont Geological Observatory, Columbia University, May 1962.

Larson, R.P. and R.D. Oldham (Argonne National Laboratory), "Plutonium in Drinking Water: Effects of Chlorination on its Maximum Permissible Concentration," Science, September 15, 1978.

Mattson, Roger J. (Director, Division of Siting, Health and Safeguards Standards, U.S. Nuclear Regulatory Commission). Statement presented at U.S. Environmental Protection Agency Hearing on Proposed 40 CFR Part 190, March 8, 1976.

Morgan, Karl Z., "Cancer and Low Level Ionizing Radiation," Bulletin of Atomic Scientists, September 1978.

National Academy of Sciences, Report of the Advisory Committee on the Biological Effects of Ionizing Radiation, The Effects on Populations of Exposure to Low Levels of Ionizing Radiation, November 1972.

Ng, Yook C., C. Ann Burton, Stanley E. Thompson, Robert K. Tandy, Helen K. Kretner, and Michael W. Pratt, Prediction of the Maximum Dosage to Man from the Fallout of Nuclear Devices: IV. Handbook for Estimating the Maximum Internal Dose from Radionuclides Released to the Biosphere (TID-4500, UC-48, Biology and Medicine). Livermore, California: Lawrence Radiation Laboratory, May 14, 1968.

"On Wisconsin," The Milwaukee Journal, October 29, 1978.

Proceedings of a Congressional Seminar on Low-Level Ionizing Radiation (A Report Transmitted by the Subcommittee on Energy and the Environment of the Committee on Interior and Insular Affairs of the U.S. House of Representatives, Ninety-fourth Congress, Second Session, November 1976). Washington, D.C.: U.S. Government Printing Office, 1976.

"Radiation Roulette? Exposure Limits: Studies Cast Doubt on Adequacy," The Day (New London, CT), June 30, 1978.

Reinig, William C., ed. Environmental Surveillance in the Vicinity of Nuclear Facilities (Proceedings of a Symposium Sponsored by the Health Physics Society). Charles C. Thomas Publishers, 1970.

Russell, R. Scott, ed. Radioactivity and the Human Diet. Oxford, New York: Pergamon Press, 1966.

"Snow Pollution is Health Danger," Wisconsin State Journal (Madison), December 29, 1974.

"State Deer Harvest Could Be Biggest Ever," The Pointer (Stevens Point, WI), November 16, 1978.

Sternglass, Ernest J. "Cancer Mortality Changes Around Nuclear Facilities in Connecticut," February 10, 1978.

Tamplin, Arthur R. "Random Violence in the Pacific," Bulletin of the Atomic Scientists, March 1974.

United Nations Food and Agriculture Organization, FAO Atomic Energy Series, No. 5, "Agricultural and Public Health Aspects of Radioactive Contamination in Normal and Emergency Situations," Papers presented at the seminar, Scheveningen, the Netherlands, 11-15 December 1961 (FAO, Rome: 1964).

U.S. Atomic Energy Commission. Concluding Statement of Position of the Regulatory Staff — Public Rulemaking Hearing on: Numerical Guides for Design Objectives and Limiting Conditions for Operation to Meet the Criterion "AS LOW AS PRACTICABLE" for Radioactive Material in Light-Water-Cooled Nuclear Power Reactors (Docket No. RM-50-2), February 20, 1974.

______________. Fallout from Nuclear Tests, Revised Ed., June 1966.

______________. Final Environmental Statement Concerning Proposed Rule Making Action: Numerical Guides for Design Objectives and Limiting Conditions for Operation to Meet the Criterion "AS LOW AS PRACTICABLE" for Radioactive Material in

———. Light-Water-Cooled Nuclear Power Reactor Effluents (WASH-1258), Volume 3, July 1973.

"USDA Would Relax Boneless Meat Rules," Wisconsin Rapids (WI) Daily Tribune, October 6, 1977.

U.S. Department of Health, Education and Welfare. Proceedings of the 4th Annual National Conference on Radiation Control: "Save A Rad." April 30 – May 4, 1972.

———. Preliminary Radiological Surveillance Studies at a Boiling Water Nuclear Power Reactor. December 1969.

———. Radiological Health Data Reports. December 1962; February 1963; November 1964.

U.S. Environmental Protection Agency, Assessment of Fallout in the United States from Atmospheric Nuclear Testing on September 26 and November 17, 1976 by the People's Republic of China, August 1977. (EPA-520/577-002).

———. Environmental Radiation Dose Commitment: An Application to the Nuclear Power Industry, February 1974.

———. Environmental Radioactivity Surveillance Guide (ORP/SID 72-2) June 1972.

———. Office of Water Supply. National Interim Primary Drinking Water Regulations (EPA-570/9-76-003), September 1976.

———. Radiation Data and Reports, Vol. 15, No. 10, October 1974.

———. Radiological Quality of the Environment in the United States, 1977 (EPA 520/1-77-009), September 1977.

U.S. Federal Register, August 14, 1975.

U.S. Nuclear Regulatory Commission. Draft Environmental Statement Related to Operation of the LaCrosse Boiling Water Reactor (NUREG-0226 – Docket No. 50-409) June 1976.

———. Final Environmental Statement Related to Construction of Tyrone Energy Park (Docket No. STN 50-484), April 1977.

———. Regulatory Guide 1.109 (Revision 1): Calculation of Annual Doses to Man from Routine Releases of Reactor Effluents for the Purpose of Evaluating Compliance with 10 CFR Part 50, Appendix I, October 1977.

Wisconsin Department of Health and Social Services, Division of Health, Section of Radiation Protection, Annual Report: Environmental Radiation Survey – LaCrosse Boiling Water Reactor for the period May 1965 through December 1976.

______________. Report of radioactive concentrations in milk samples received by League Against Nuclear Dangers, Inc. (Stevens Point, WI) and Public Document Room, University of Wisconsin – Stevens Point Learning Resources Center, March 1975.

______________. Computer printouts of radionuclide concentrations in milk samples for six non-nuclear testing stations, 1963 through 1976.

Wisconsin Legislative Reference Bureau, The State of Wisconsin 1975 Blue Book. Madison, WI: Department of Administration, Document Sales and Distribution, 1975.